Praise for John Horgan's *The End of Science*

"In this **wonderful, provoca**[...] [...] to take us
along while he buttonho[...] [...] most
opinionated, most exasperati[...] [...] views on where science
is and where it's going. . . . T[...] all come to life in Horgan's narrative."
—*Washington Post Book World*, front page review

"Thanks to Mr. Horgan's smooth prose style, puckish sense of humor,
and wicked eye for detail, these encounters make for zesty reading.
Frequently they are hilarious. . . . **A thumping good book.**"
—*Wall Street Journal*

"**A deft wordsmith** and **keen observer,** Horgan offers lucid expositions of
everything from superstring theory and Thomas Kuhn's analysis of
scientific revolutions to the origin of life and sociobiology."
—*Business Week*

". . . [The book's] greatest pleasures flow from Horgan's encounters with
the characters who have made science their lives."
—*San Francisco Chronicle*

"Rich in **provocative** ideas and **insightful** anecdotes . . ."
—*Library Journal*

"The *End of Science*, a lively, witty book by science writer John Horgan,
exemplifies the genre's virtues and faults: **It is fun to read. It will make
people think.**"
—*Reason*

"**It is a sweeping argument, admirably brought off . . .**"
—*Washington Times*

"**John Horgan has everybody talking.** Probably no science book of this
year has generated as much comment."
—*Rocky Mountain News*

"*The End of Science* is a revealing glimpse into the minds of some of our
leading scientists and philosophers. **Read it. Enjoy it, learn from it.**"
—*Hartford Courant*

THE END OF SCIENCE

THE
END
OF
SCIENCE

Facing the Limits of Knowledge
in the Twilight of the Scientific Age

JOHN HORGAN

BROADWAY BOOKS

NEW YORK

BROADWAY

A hardcover edition of this book was originally published in 1996 by Addison-Wesley Publishing Company, Inc. It is here reprinted by arrangement with Addison-Wesley Publishing Company, Inc.

First Broadway Books trade paperback edition published 1997.

Library of Congress Cataloging-in-Publication Data

Horgan, John, 1953–
    The end of science : facing the limits of knowledge in the
twilight of the scientific age / John Horgan.
        p.   cm. — (Helix books)
    Originally published: Reading, Mass. : Addison-Wesley Pub., c1996.
    Includes bibliographical references and index.
    ISBN 0-553-06174-7 (pbk.)
    1. Science—Philosophy.   2. Science—History.
Q175.H794   1997   501—dc21   97-1185
                                                    CIP

Text design by Diane Levy

97 98 99 00 01 10 9 8 7 6 5 4 3 2 1

For Suzie

Contents

Is Daniel Dennett a mysterian?
Marvin Minsky's fear of single-mindedness.
The triumph of materialism.

Searching for
The Answer

It was in the summer of 1989, during a trip to upstate New York, that I began to think seriously about the possibility that science, pure science, might be over. I had flown to the University of Syracuse to interview Roger Penrose, a British physicist who was a visiting scholar there. Before meeting Penrose, I had struggled through galleys of his dense, difficult book, *The Emperor's New Mind*, which to my astonishment became a best-seller several months later, after being praised in the *New York Times Book Review*.[1] In the book, Penrose cast his eye across the vast panorama of modern science and found it wanting. This knowledge, Penrose asserted, for all its power and richness, could not possibly account for the ultimate mystery of existence, human consciousness.

The key to consciousness, Penrose speculated, might be hidden in the fissure between the two major theories of modern physics: quantum mechanics, which describes electromagnetism and the nuclear forces, and general relativity, Einstein's theory of gravity. Many physicists, beginning with Einstein, had tried and failed to fuse quantum mechanics and general relativity into a single, seamless "unified" theory. In his book, Penrose sketched out what a unified theory might look like and how it might give rise to thought. His scheme, which involved exotic quantum and gravitational effects percolating through the brain, was vague, convoluted, utterly unsupported by evidence from physics or neuroscience. But if it turned out to be in any sense right, it would represent a monumental achievement, a theory that in one stroke would unify physics and solve one of philosophy's most vexing problems, the link between mind and matter. Penrose's ambition alone, I thought, would make him an excellent subject for a profile in *Scientific American*, which employed me as a staff writer.[2]

When I arrived at the airport in Syracuse, Penrose was waiting for me. He was an elfin man, capped with a shock of black hair, who seemed simultaneously distracted and acutely alert. As he drove us back to the Syracuse campus, he kept wondering aloud if he was going in the right direction. He seemed awash in mysteries. I found myself in the disconcerting position of recommending that he take this exit, or make that turn, although I had never been in Syracuse before. In spite of our combined ignorance, we managed to make our way without incident to the building where Penrose worked. On entering Penrose's office we discovered that a colleague had left a brightly colored aerosol can labeled Superstring on his desk. When Penrose pushed the button on the top of the can, a lime green, spaghetti-like strand shot across the room.

Penrose smiled at this little insider's joke. Superstring is the name not only of a child's toy, but also of an extremely small and extremely hypothetical stringlike particle posited by a popular theory of physics. According to the theory, the wriggling of these strings in a 10-dimensional hyperspace generates all the matter and energy in the universe and even space and time. Many of the world's leading physicists felt that superstring theory might turn out to be the unified theory they had sought for so long; some even called it a theory of everything. Penrose was not among the faithful. "It couldn't be right," he told me. "It's just not the way I'd expect the answer to be." I began to realize, as Penrose spoke, that to him "the answer" was more than a mere theory of physics, a way of organizing data and predicting events. He was talking about *The Answer*: the secret of life, the solution to the riddle of the universe.

Penrose is an admitted Platonist. Scientists do not invent the truth; they discover it. Genuine truths exude a beauty, a rightness, a self-evident quality that gives them the power of revelation. Superstring theory did not possess these traits, in Penrose's mind. He conceded that the "suggestion" he set forth in *The Emperor's New Mind*—it did not merit the term *theory* yet, he admitted—was rather ungainly. It might turn out to be wrong, certainly in its details. But he felt sure that it was closer to the truth than was superstring theory. In saying that, I asked, was Penrose implying that one day scientists would find *The Answer* and thus bring their quest to an end?

Unlike some prominent scientists, who seem to equate tentativity with weakness, Penrose actually thinks before he responds, and even as he responds. "I don't think we're close," he said slowly, squinting out his office window, "but it doesn't mean things couldn't move fast at some stage." He cogitated some more. "I guess this is rather suggesting that there *is* an answer," he continued, "although perhaps that's too pessimistic." This final comment stopped me short. What is so pessimistic, I asked, about a truth seeker thinking that the truth is attainable? "Solving mysteries is a wonderful thing to do," Penrose replied. "And if they were all solved, somehow, that would be rather boring." Then he chuckled, as if struck by the oddness of his own words.[3]

Long after leaving Syracuse, I mulled over Penrose's remarks. Was it possible that science could come to an end? Could scientists, in effect, learn everything there is to know? Could they banish mystery from the universe? It was hard for me to imagine a world without science, and not only because my job depended on it. I had become a science writer in large part because I considered science—pure science, the search for knowledge for its own sake—to be the noblest and most meaningful of human endeavors. We are here to figure out why we are here. What other purpose is worthy of us?

I had not always been so enamored of science. In college, I passed through a phase during which literary criticism struck me as the most thrilling of intellectual endeavors. Late one night, however, after too many cups of coffee, too many hours spent slogging through yet another interpretation of James Joyce's *Ulysses*, I had a crisis of faith. Very smart people had been arguing for decades over the meaning of *Ulysses*. But one of the messages of modern criticism, and of modern literature, was that all texts are "ironic": they have multiple meanings, none of them definitive.[4] *Oedipus Rex, The Inferno*, even the Bible are in a sense "just kidding," not to be taken too literally. Arguments over meaning can never be resolved, since the only true meaning of a text is the text itself. Of course, this message applied to the critics, too. One was left with an infinite regress of interpretations, none of which represented the final word. But everyone still kept arguing! To what end? For each critic to be more clever, more *interesting*, than the rest? It all began to seem pointless.

Although I was an English major, I took at least one course in science or mathematics every semester. Working on a problem in calculus or physics represented a pleasant change of pace from messy humanities assignments; I found great satisfaction in arriving at the correct answer to a problem. The more frustrated I became with the ironic outlook of literature and literary criticism, the more I began to appreciate the crisp, no-nonsense approach of science. Scientists have the ability to pose questions and resolve them in a way that critics, philosophers, historians cannot. Theories are tested experimentally, compared to reality, and those found wanting are rejected. The power of science cannot be denied: it has given us computers and jets, vaccines and thermonuclear bombs, technologies that, for better or worse, have altered the course of history. Science, more than any other mode of knowledge—literary criticism, philosophy, art, religion—yields durable insights into the nature of things. It gets us somewhere. My mini-epiphany led, eventually, to my becoming a science writer. It also left me with this criterion for science: science addresses questions that can be answered, at least in principle, given a reasonable amount of time and resources.

Before my meeting with Penrose, I had taken it for granted that science was open-ended, even infinite. The possibility that scientists might one day find a truth so potent that it would obviate all further investigations had struck me as wishful thinking at best, or as the kind of hyperbole required to sell science (and science books) to the masses. The earnestness, and ambivalence, with which Penrose contemplated the prospect of a final theory forced me to reassess my own views of science's future. Over time, I became obsessed with the issue. What are the limits of science, if any? Is science infinite, or is it as mortal as we are? If the latter, is the end in sight? Is it upon us?

After my original conversation with Penrose, I sought out other scientists who were butting their heads against the limits of knowledge: particle physicists who dreamed of a final theory of matter and energy; cosmologists trying to understand precisely how and even why our universe was created; evolutionary biologists seeking to determine how life began and what laws governed its subsequent unfolding; neuroscientists probing the processes in the brain that give rise to consciousness; explorers of chaos and complexity, who hoped that with computers and new mathematical techniques they could revitalize science. I also spoke to philosophers, including some who allegedly doubted whether science could ever achieve objective, absolute truths. I wrote articles about a number of these scientists and philosophers for *Scientific American*.

When I first thought about writing a book, I envisioned it as a series of portraits, warts and all, of the fascinating truth seekers and truth shunners I have been fortunate enough to interview. I intended to leave it to readers to decide whose forecasts about the future of science made sense and whose did not. After all, who really knew what the ultimate limits of knowledge might be? But gradually, I began to imagine that *I* knew; I convinced myself that one particular scenario was more plausible than all the others. I decided to abandon any pretense of journalistic objectivity and write a book that was overtly judgmental, argumentative, and personal. While still focusing on individual scientists and philosophers, the book would present my views as well. That approach, I felt, would be more in keeping with my conviction that most assertions about the limits of knowledge are, finally, deeply idiosyncratic.

It has become a truism by now that scientists are not mere knowledge-acquisition machines; they are guided by emotion and intuition as well as by cold reason and calculation. Scientists are rarely so human, I have found, so at the mercy of their fears and desires, as when they are confronting the limits of knowledge. The greatest scientists want, above all, to discover truths about nature (in addition to acquiring glory, grants, and tenure and improving the lot of humankind); they want to *know*. They hope, and trust, that the truth is attainable, not merely an ideal or asymptote, which they eternally approach. They also believe, as I do, that the quest for knowledge is by far the noblest and most meaningful of all human activities.

Scientists who harbor this belief are often accused of arrogance. Some *are* arrogant, supremely so. But many others, I have found, are less arrogant than anxious. These are trying times for truth seekers. The scientific enterprise is threatened by technophobes, animal-rights activists, religious fundamentalists, and, most important, stingy politicians. Social, political, and economic constraints will make it more difficult to practice science, and pure science in particular, in the future.

Moreover, science itself, as it advances, keeps imposing limits on its own power. Einstein's theory of special relativity prohibits the transmission of matter or even information at speeds faster than that of light; quantum mechanics dictates that our knowledge of the microrealm will always be uncertain; chaos theory confirms that even without quantum indeterminacy many phenomena would be impossible to predict; Kurt Gödel's incompleteness theorem denies us the possibility of constructing a complete, consistent mathematical description of reality. And evolutionary biology keeps reminding us that we are animals, designed by natural selection not for discovering deep truths of nature, but for breeding.

Optimists who think they can overcome all these limits must face yet another quandary, perhaps the most disturbing of all. What will scientists do if they succeed in knowing what can be known? What, then, would be the purpose of life? What would be the purpose of humanity? Roger Penrose revealed his anxiety over this dilemma when he called his dream of a final theory pessimistic.

Given these troubling issues, it is no wonder that many scientists whom I interviewed for this book seemed gripped by a profound unease. But their malaise, I will argue, has another, much more immediate cause. *If one believes in science*, one must accept the possibility—even the probability—that the great era of scientific discovery is over. By *science* I mean not applied science, but science at its purest and grandest, the primordial human quest to understand the universe and our place in it. Further research may yield no more great revelations or revolutions, but only incremental, diminishing returns.

The Anxiety of Scientific Influence

In trying to understand the mood of modern scientists, I have found that ideas from literary criticism can serve some purpose after all. In his influential 1973 essay, *The Anxiety of Influence*, Harold Bloom likened the modern poet to Satan in Milton's *Paradise Lost*.[5] Just as Satan fought to assert his individuality by defying the perfection of God, so must the modern poet engage in an Oedipal struggle to define himself or herself in relation to Shakespeare, Dante, and other masters. The effort is ultimately futile, Bloom said, because no poet can hope to approach, let alone surpass, the perfection of such forebears. Modern poets are all essentially tragic figures, latecomers.

Modern scientists, too, are latecomers, and their burden is much heavier than that of poets. Scientists must endure not merely Shakespeare's *King Lear*, but Newton's laws of motion, Darwin's theory of natural selection, and Einstein's theory of general relativity. These theories are not merely beautiful; they are also true, empirically true, in a way that no work of art can be. Most researchers simply concede their inability to supersede what Bloom called "the embarrassments of a tradition grown too wealthy to need anything more."[6] They try to solve what philosopher of science Thomas Kuhn has patronizingly called "puzzles," problems whose solution buttresses the prevailing paradigm. They settle for refining and applying the brilliant, pioneering discoveries of their predecessors. They try to measure the mass of quarks more precisely, or to determine how a given stretch of DNA guides the growth of the embryonic brain. Others become what Bloom derided as a "mere rebel, a childish inverter of conventional moral categories."[7] The rebels denigrate the dominant theories of science as flimsy social fabrications rather than rigorously tested descriptions of nature.

Bloom's "strong poets" accept the perfection of their predecessors and yet strive to transcend it through various subterfuges, including a subtle misreading of the predecessors' work; only by so doing can modern poets break free of the stultifying influence of the past. There are strong scientists, too, those who are seeking to misread and therefore to transcend quantum mechanics or the big bang theory or Darwinian evolution. Roger Penrose is a strong scientist. For the most part, he and others of his ilk have only one option: to pursue science in a speculative, postempirical mode that I call ironic science. Ironic science resembles literary criticism in that it offers points of view, opinions, which are, at best, interesting, which provoke further comment. But it does not converge on the truth. It cannot achieve empirically verifiable surprises that force scientists to make substantial revisions in their basic description of reality.

The most common strategy of the strong scientist is to point to all the shortcomings of current scientific knowledge, to all the questions left unanswered. But the questions tend to be ones that may *never* be definitively answered given the limits of human science. How, exactly, was the universe created? Could our universe be just one of an infinite number of universes? Could quarks and electrons be composed of still smaller particles, ad infinitum? What does quantum mechanics really mean? (Most questions concerning meaning can only be answered ironically, as literary critics know.) Biology has its own slew of insoluble riddles. How, exactly, did life begin on earth? Just how inevitable was life's origin and its subsequent history?

The practitioner of ironic science enjoys one obvious advantage over the strong poet: the appetite of the reading public for scientific "revolutions." As empirical science ossifies, journalists such as myself, who feed society's hunger, will come under more pressure to tout theories that supposedly transcend quantum mechanics or the big bang theory or natural selection. Journalists are, after all, largely responsible for the popular impression that fields such as chaos and complexity represent genuinely new sciences superior to the stodgy old reductionist methods of Newton, Einstein, and Darwin. Journalists, myself included, have also helped Roger Penrose's ideas about consciousness win an audience much larger than they deserve given their poor standing among professional neuroscientists.

I do not mean to imply that ironic science has no value. Far from it. At its best ironic science, like great art or philosophy or, yes, literary criticism, induces wonder in us; it keeps us in awe before the mystery of the universe. But it cannot achieve its goal of transcending the truth we already have. And it certainly cannot give us—in fact, it protects us from—*The Answer*, a truth so potent that it quenches our curiosity once and for all time. After all, science itself decrees that we humans must always be content with partial truths.

Through most of this book, I will examine science as it is practiced today, by humans. (Chapter 2 takes up philosophy.) In the final two chapters I will consider the possibility—advanced by a surprising number of scientists and philosophers—that one day we humans will create intelligent machines that can transcend our puny knowledge. In my favorite version of this scenario, machines transform the entire cosmos into a vast, unified, information-processing network. All matter becomes mind. This proposal is not science, of course, but wishful thinking. It nonetheless raises some interesting questions, questions normally left to theologians. What would an all-powerful cosmic computer do? What would it think about? I can imagine only one possibility. It would try to answer *The Question*, the one that lurks behind all other questions, like an actor playing all the parts of a play: Why is there something rather than nothing? In its effort to find *The Answer* to *The Question*, the universal mind may discover the ultimate limits of knowledge.

CHAPTER ONE

The End of Progress

In 1989, just a month after my meeting with Roger Penrose in Syracuse, Gustavus Adolphus College in Minnesota held a symposium with the provocative but misleading title, "The End of Science?" The meeting's premise was that *belief* in science—rather than science itself—was coming to an end. As one organizer put it, "There is an increasing feeling that science as a unified, universal, objective endeavor is over."[1] Most of the speakers were philosophers who had challenged the authority of science in one way or another. The meeting's great irony was that one of the scientists who spoke, Gunther Stent, a biologist at the University of California at Berkeley, had for years promulgated a much more dramatic scenario than the one posited by the symposium. Stent had asserted that science itself might be ending, and not because of the skepticism of a few academic sophists. Quite the contrary. Science might be ending because it worked so well.

Stent is hardly a fringe figure. He was a pioneer of molecular biology; he founded the first department dedicated to that field at Berkeley in the 1950s and performed experiments that helped to illuminate the machinery of genetic transmission. Later, after switching from genetics to the study of the brain, he was named chairman of the neurobiology department of the National Academy of Sciences. Stent is also the most astute analyst of the limits of science whom I have encountered (and by astute I mean of course that he articulates my own inchoate premonitions). In the late 1960s, while Berkeley was racked with student protests, he wrote an astonishingly prescient book, now long out of print, called *The Coming of the Golden Age: A View of the End of Progress*. Published in 1969, it contended that science—as well as technology, the arts, and all progressive, cumulative enterprises—was coming to an end.[2]

Most people, Stent acknowledged, consider the notion that science might soon cease to be absurd. How can science possibly be nearing an end when it has been advancing so rapidly throughout this century? Stent turned this inductive argument on its head. Initially, he granted, science advances exponentially through a positive feedback effect; knowledge begets more knowledge, and power begets more power. Stent credited the American historian Henry Adams with having foreseen this aspect of science at the turn of the century.[3]

Adams's law of acceleration, Stent pointed out, has an interesting corollary. If there are any limits to science, any barriers to further progress, then science may well be moving at unprecedented speed just before it crashes into them. When science seems most muscular, triumphant, potent, that may be when it is nearest death. "Indeed, the dizzy rate at which progress is now proceeding," Stent wrote in *Golden Age*, "makes it seem very likely that progress must come to a stop soon, perhaps in our lifetime, perhaps in a generation or two."[4]

Certain fields of science, Stent argued, are limited simply by the boundedness of their subject matter. No one would consider human anatomy or geography, for example, to be infinite endeavors. Chemistry, too, is bounded. "[T]hough the total number of possible chemical reactions is very great and the variety of reactions they can undergo vast, the goal of chemistry of understanding the principles governing the behavior of such molecules is, like the goal of geography, clearly limited."[5] That goal, arguably, was achieved in the 1930s, when the chemist Linus Pauling showed how all chemical interactions could be understood in terms of quantum mechanics.[6]

In his own field of biology, Stent asserted, the discovery of DNA's twin-corkscrew structure in 1953 and the subsequent deciphering of the genetic code had solved the profound problem of how genetic information is passed on from one generation to the next. Biologists had only three major questions left to explore: how life began, how a single fertilized cell develops into a multicellular organism, and how the central nervous system processes information. When those goals are achieved, Stent said, the basic task of biology, pure biology, will be completed.

Stent acknowledged that biologists could, in principle, continue exploring specific phenomena and applying their knowledge forever. But according to Darwinian theory, science stems not from our desire for truth per se, but from our compulsion to control our environment in order to increase the likelihood that our genes will propagate. When a given field of science begins to yield diminishing practical returns, scientists may have less incentive to pursue their research and society may be less inclined to pay for it.

Moreover, just because biologists complete their empirical investigations, Stent asserted, does not mean that they will have answered all relevant questions. For example, no purely physiological theory can ever really *explain* consciousness, since the "processes responsible for this wholly private experience will be seen to degenerate into seemingly quite ordinary, workaday reactions, no more or less fascinating than those that occur in, say, the liver. . . ."[7]

Unlike biology, Stent wrote, the physical sciences seem to be open-ended. Physicists can always attempt to probe more deeply into matter by smashing particles against each other with greater force, and astronomers can always strive to see further into the universe. But in their efforts to gather data from ever-more-remote regimes, physicists will inevitably confront various physical, economic, and even cognitive limits.

Over the course of this century, physics has become more and more difficult to comprehend; it has outrun our Darwinian epistemology, our innate concepts for coping with the world. Stent rejected the old argument that "yesterday's nonsense is today's common sense."[8] Society may be willing to support continued research in physics as long as it has the potential to generate powerful new technologies, such as nuclear weapons and nuclear power. But when physics becomes impractical as well as incomprehensible, Stent predicted, society will surely withdraw its support.

Stent's prognosis for the future was an odd mixture of optimism and pessimism. He predicted that science, before it ends, might help to solve many of civilization's most pressing problems. It could eliminate disease and poverty and provide society with cheap, pollution-free energy, perhaps through the harnessing of fusion reactions. As we gain more dominion over nature, however, we may lose what Nietzsche called our "will to power"; we may become less motivated to pursue further research—especially if such research has little chance of yielding tangible benefits.

As society becomes more affluent and comfortable, fewer young people may choose the increasingly difficult path of science or even of the arts. Many may turn to more hedonistic pursuits, perhaps even abandoning the real world for fantasies induced by drugs or electronic devices feeding directly into the brain. Stent concluded that sooner or later, progress would "stop dead in its tracks," leaving the world in a largely static condition that he called "the new Polynesia." The advent of beatniks and hippies, he surmised, signaled the beginning of the end of progress and the dawn of the new Polynesia. He closed his book with the sardonic comment that "millennia of doing arts and sciences will finally transform the tragicomedy of life into a happening."[9]

A Trip to Berkeley

In the spring of 1992 I traveled to Berkeley to see how Stent thought his predictions had held up over the years.[10] Strolling toward the university from my hotel, I passed what appeared to be the detritus of the sixties: men and women with long gray hair and ragged clothes asking for spare change. Once on the campus, I made my way to the university's biology building, a hulking, concrete complex shadowed by dusty eucalyptus trees. I took an elevator one floor up to Stent's laboratory and found it locked. A few minutes later the elevator door slid open and out walked Stent, a red-faced, sweaty man wearing a yellow bicycle helmet and rolling a dirt-encrusted mountain bike.

Stent had moved to the United States from Germany as a youth, and his gruff voice and attire still bore traces of his origins. He wore wire-rimmed glasses, a blue, short-sleeved shirt with epaulets, dark slacks, and shiny black shoes. He led me through his laboratory, crammed with microscopes, centrifuges, and scientific glassware, to a small office at the rear. The hall outside his office was adorned with photographs and paintings of Buddha. When Stent closed the door of his office behind us, I saw that he had tacked to the door's inner surface a poster from the 1989 meeting at Gustavus Adolphus College. The top half of the poster was covered with the word SCIENCE, written in huge, luridly colored letters. The letters were melting, oozing downward into a pool of Day-Glo protoplasm. Beneath this psychedelic puddle big black letters asked, "The End of Science?"

Stent, at the beginning of our interview, seemed rather suspicious. He asked pointedly if I was following the legal travails of the journalist Janet Malcolm, who had just lost a round in her interminable legal battle with a former profile subject, the psychoanalyst Jeffrey Masson. I mumbled something to the effect that Malcolm's trangressions were too minor to merit any punishment, but that her methods did seem rather careless. If I were writing something critical about a person as obviously volatile as Masson, I told Stent, I would be sure to have all my quotes on tape. (As I spoke, my own tape recorder was silently spinning between us.)

Gradually, Stent relaxed and began to tell me about his life. Born in Berlin to Jewish parents in 1924, he escaped from Germany in 1938 and moved in with a sister living in Chicago. He obtained a doctorate in chemistry at the University of Illinois, but upon reading Erwin Schrödinger's book *What Is Life?* he became entranced by the mystery of genetic transmission. After working at the California Institute of Technology with the eminent biophysicist Max Delbrück, Stent obtained a professorship at Berkeley in 1952. In these early years of molecular biology, Stent said, "none of us knew what we were doing. Then Watson and Crick found the double helix, and within a few weeks we realized we were doing molecular biology."

Stent began pondering the limits of science in the 1960s, partly in reaction to Berkeley's free-speech movement, which challenged the value of Western rationalism and technological progress and other aspects of civilization that Stent held dear. The university appointed him to a committee to "deal with this, to calm things down," by talking to students. Stent sought to fulfill this mandate—and to resolve his own inner conflicts over his role as a scientist—by delivering a series of lectures. These lectures became *The Coming of the Golden Age.*

I told Stent that I could not determine, after finishing *The Coming of the Golden Age*, whether he believed that the new Polynesia, the era of social and intellectual stasis and universal leisure, would be an improvement over our present situation. "I could never decide this!" he exclaimed, looking genuinely distressed. "People called me a pessimist, but I thought I was an optimist." He certainly did not think such a society would be in any sense utopian. After the horrors wreaked by totalitarian states in this century, he explained, it was no longer possible to take the idea of utopia seriously.

Stent felt that his predictions had held up reasonably well. Although hippies had vanished (except for the pitiful relics on Berkeley's streets),

American culture had become increasingly materialistic and anti-intellectual; hippies had evolved into yuppies. The cold war had ended, although not through the gradual merging of communist and capitalist states that Stent had envisioned. He admitted that he had not anticipated the resurgence, in the wake of the cold war, of long-repressed ethnic conflicts. "I'm very depressed at what's happening in the Balkans," he said. "I didn't think that would happen." Stent was also surprised by the persistence of poverty and of racial conflict in the United States, but he believed these problems would eventually diminish in importance. (Aha, I thought. He was an optimist after all.)

Stent was convinced that science was showing signs of the closure he had predicted in *Golden Age*. Particle physicists were having difficulty convincing society to pay for their increasingly expensive experiments, such as the superconducting supercollider. As for biologists, they still had much to learn about how, say, a fertilized cell is transformed into a complex, multicellular organism, such as an elephant, and about the workings of the brain. "But I think the big picture is basically over," he said. Evolutionary biology in particular "was over when Darwin published *The Origin of Species*," Stent said. He scoffed at the hope of some evolutionary biologists—notably Edward Wilson of Harvard—that they could remain occupied indefinitely by doing a thorough survey of all life on earth, species by species. Such an enterprise would be a mindless "glass bead game," Stent complained.

He then plunged into a diatribe against environmentalism. It was at heart an antihuman philosophy, one that contributed to the low self-esteem of American youth and poor black children in particular. Alarmed that my favorite Cassandra was revealing himself to be a crank, I changed the subject to consciousness. Did Stent still consider consciousness to be an unsolvable scientific problem, as he had suggested in *Golden Age*? He replied that he thought very highly of Francis Crick, who late in his career had turned his attention to consciousness. If Crick felt that consciousness was scientifically tractable, Stent said, then that possibility must be taken seriously.

Stent was still convinced, though, that a purely physiological explanation of consciousness would not be as comprehensible or as meaningful as most people would like, nor would it help us to solve moral and ethical questions. Stent thought the progress of science might give religion a clearer role in the future rather than eliminate it entirely, as many scientists had once hoped. Although it could not compete with science's far more compelling stories about the physical realm, religion retained some value in offering moral guidance. "Humans are animals, but we're also moral subjects. The task of religion is more and more in the moral realm."

When I asked about the possibility that computers might become intelligent and create their own science, Stent snorted in derision. He had a dim view of artificial intelligence, and particularly of its more visionary enthusiasts. Computers may excel at precisely defined tasks such as mathematics and chess, he pointed out, but they still perform abysmally when confronted with the kind of problems—recognizing a face or a voice or walking down a crowded sidewalk—that humans solve effortlessly. "They're full of it," Stent said of Marvin Minsky and others who have predicted that one day we humans will be able to download our personalities into computers. "I wouldn't rule out the possibility that in the twenty-third century you might have an artificial brain," he added, "but it would need experience." One could design a computer to become an expert in restaurants, "but this machine would never know what a steak tastes like."

Stent was similarly skeptical of the claims of investigators of chaos and complexity that with computers and sophisticated mathematics they would be able to transcend the science of the past. In *The Coming of the Golden Age*, Stent discussed the work of one of the pioneers of chaos theory, Benoit Mandelbrot. Beginning in the early 1960s, Mandelbrot showed that many phenomena are intrinsically indeterministic: they exhibit behavior that is unpredictable and apparently random. Scientists can only guess at the causes of individual events and cannot predict them with any accuracy.

Investigators of chaos and complexity were attempting to create effective, comprehensible theories of the same phenomena studied by Mandelbrot, Stent said. He had concluded in *Golden Age* that these indeterministic phenomena would resist scientific analysis, and he saw no reason to change that assessment. Quite the contrary. The work emerging from those fields demonstrated his point that science, when pushed too far, always culminates in incoherence. So Stent did not think that chaos and complexity would bring about the rebirth of science? "No," he replied with a rakish grin. "It's the end of science."

What Science Has Accomplished

We obviously are nowhere near the new Polynesia that Stent envisioned, in part because applied science has not come nearly as far as Stent had hoped (feared?) when he wrote *The Coming of the Golden Age*. But I have come to the conclusion that Stent's prophecy has, in one very important sense, already come to pass. Pure science, the quest for knowledge about what we are and where we came from, has already entered an era of diminishing returns. By far the greatest barrier to future progress in pure science is its past success. Researchers have already mapped out physical reality, ranging from the microrealm of quarks and electrons to the macrorealm of planets, stars, and galaxies. Physicists have shown that all matter is ruled by a few basic forces: gravity, electromagnetism, and the strong and weak nuclear forces.

Scientists have also stitched their knowledge into an impressive, if not terribly detailed, narrative of how we came to be. The universe exploded into existence 15 billion years ago, give or take 5 billion years (astronomers may never agree on an exact figure), and is still expanding outward. Some 4.5 billion years ago, the detritus of an exploding star, a supernova, condensed into our solar system. Sometime during the next few hundred million years, for reasons that may never be known, single-celled organisms bearing an ingenious molecule called DNA emerged on the still-hellish earth. These Adamic microbes gave rise, by means of natural selection, to an extraordinary array of more complex creatures, including *Homo sapiens*.

My guess is that this narrative that scientists have woven from their knowledge, this modern myth of creation, will be as viable 100 or even 1,000 years from now as it is today. Why? Because it is true. Moreover, given how far science has already come, and given the physical, social, and cognitive limits constraining further research, science is unlikely to make any significant additions to the knowledge it has already generated. There will be no great revelations in the future comparable to those bestowed upon us by Darwin or Einstein or Watson and Crick.

The Anticlimax of Immortality

Applied science will continue for a long time to come. Scientists will keep developing versatile new materials; faster and more sophisticated computers; genetic-engineering techniques that make us healthier, stronger, longer-lived; perhaps even fusion reactors that provide cheap energy with few environmental side effects (although given the drastic cutbacks in funding, fusion's prospects now seem dimmer than ever). The question is, will these advances in applied science bring about any surprises, any revolutionary shifts in our basic knowledge? Will they force scientists to revise the map they have drawn of the universe's structure or the narrative they have constructed of our cosmic creation and history? Probably not. Applied science in this century has tended to reinforce rather than to challenge the prevailing theoretical paradigms. Lasers and transistors confirm the power of quantum mechanics, just as genetic engineering bolsters belief in the DNA-based model of evolution.

What constitutes a surprise? Einstein's discovery that time and space, the I beams of reality, are made of rubber was a surprise. So was the observation by astronomers that the universe is expanding, evolving. Quantum mechanics, which unveiled a probabilistic element, a Lucretian swerve, at the bottom of things, was an enormous surprise; God *does* play dice (Einstein's disapproval notwithstanding). The later finding that protons and neutrons are made of smaller particles called quarks was a much lesser surprise, because it merely extended quantum theory to a deeper domain; the foundations of physics remained intact.

Learning that we humans were created not *de novo* by God, but gradually, by the process of natural selection, was a big surprise. Most other aspects of human evolution—those concerning where, when, and how, precisely, *Homo sapiens* evolved—are details. These details may be interesting, but they are not likely to be surprising unless they show that scientists' basic assumptions about evolution are wrong. We may learn, say, that our sudden surge in intelligence was catalyzed by the intervention of alien beings, as in the movie *2001*. That would be a very big surprise. In fact, any proof that life exists—or even once existed—beyond our little planet would constitute a huge surprise. Science, and all human thought, would be reborn. Speculation about the origin of life and its inevitability would be placed on a much more empirical basis.

But how likely is it that we will discover life elsewhere? In retrospect, the space programs of both the United States and the USSR represented elaborate displays of saber rattling rather than the opening of a new frontier for human knowledge. The prospects for space exploration on anything more than a trivial level seem increasingly unlikely. We no longer have the will or the money to indulge in technological muscle flexing for its own sake. Humans, made of flesh and blood, may someday travel to other planets here in our solar system. But unless we find some way to transcend Einstein's prohibition against faster-than-light travel, chances are that we will never even attempt to visit another star, let alone another galaxy. A spaceship that can travel one million miles an hour, a velocity at least one order of magnitude greater than any current technology can attain, would still take almost 3,000 years to reach our nearest stellar neighbor, Alpha Centauri.[11]

The most dramatic advance in applied science I can imagine is immortality. Many scientists are now attempting to identify the precise causes of aging. It is conceivable that if they succeed, scientists may be able to design versions of *Homo sapiens* that can live indefinitely. But immortality, although it would represent a triumph of applied science, would not necessarily change our fundamental knowledge of the universe. We would not have any better idea of why the universe came to be and of what lies beyond its borders than we do now. Moreover, evolutionary biologists suggest that immortality may be impossible to achieve. Natural selection designed us to live long enough to breed and raise our children. As a result, senescence does not stem from any single cause or even a suite of causes; it is woven inextricably into the fabric of our being.[12]

That's What They Thought 100 Years Ago

It is easy to understand why so many people find it hard to believe that science, pure or impure, might be ending. Just a century ago, no one could imagine what the future held in store. Television? Jets? Space stations? Nuclear weapons? Computers? Genetic engineering? It must be as impossible for us to know the future of science—pure or applied—as it would have been for Thomas Aquinas to anticipate Madonna or microwave ovens. There are marvels, utterly unpredictable, lying in wait for us just as there were for our ancestors. We will only fail to seize these treasures if we decide that they do not exist and cease striving to find them. The prophecy can only be self-fulfilling.

This position is often expressed as the that's-what-they-thought-at-the-end-of-the-last-century argument. The argument goes like this: As the nineteenth century wound down, physicists thought they knew everything. But no sooner had the twentieth century begun, than Einstein and other physicists discovered—invented?—relativity theory and quantum mechanics. These theories eclipsed Newtonian physics and opened up vast new vistas for modern physics and other branches of science. Moral: Anyone who predicts that science is nearing its end will surely turn out to be as shortsighted as those nineteenth-century physicists.

Those who believe science is finite have a standard retort for this argument: the earliest explorers, because they could not find the edge of the earth, might well have concluded that it was infinite, but they would have been wrong. Moreover, it is by no means a matter of historical record that late-nineteenth-century physicists felt they had wrapped things up. The best evidence for a sense of completion is a speech given in 1894 by Albert Michelson, whose experiments on the velocity of light helped to inspire Einstein's theory of special relativity.

> While it is never safe to say that the future of Physical Science has no marvels even more astonishing than those of the past, it seems probable that most of the grand underlying principles have been firmly established and that further advances are to be sought chiefly in the rigorous application of these principles to all the phenomena which come under our notice. It is here that the science of measurement shows its importance—where quantitative results are more to be desired than qualitative work. An eminent physicist has remarked that the future truths of Physical Science are to be looked for in the sixth place of decimals.[13]

Michelson's remark about the sixth place of decimals has been so widely attributed to Lord Kelvin (after whom the Kelvin, a unit of temperature, is named) that some authors simply credit him with the quote.[14] But historians have found no evidence that Kelvin made such a statement. Moreover, at the time of Michelson's remarks, physicists were vigorously debating fundamental issues, such as the viability of the atomic theory of matter, according to the historian of science Stephen Brush of the University of Maryland. Michelson was so absorbed in his optics experiments, Brush suggested, that he was "oblivious to the violent controversies raging among theorists at the time." The alleged "Victorian calm in physics," Brush concluded, is a "myth."[15]

The Apocryphal Patent Official

Other historians, predictably, disagree with Brush's assessment.[16] Questions concerning the mood of a given era can never be completely resolved, but the view that scientists in the last century were complacent about the state of their field has clearly been exaggerated. Historians have provided a definitive ruling on another anecdote favored by those reluctant to accept that science might be mortal. The story alleges that in the mid-1800s the head of the U.S. Patent Office quit his job and recommended that the office be shut down because there would soon be nothing left to invent.

In 1995, Daniel Koshland, editor of the prestigious journal *Science*, repeated this story in an introduction to a special section on the future of science. In the section, leading scientists offered predictions about what their fields might accomplish over the next 20 years. Koshland, who, like Gunther Stent, is a biologist at the University of California at Berkeley, exulted that his prognosticators "clearly do not agree with that commissioner of patents of yesteryear. Great discoveries with great import for the future of science are in the offing. That we have come so far so fast is not an indication that we have saturated the discovery market, but rather that discoveries will come even faster."[17]

There were two problems with Koshland's essay. First, the contributors to his special section envisioned not "great discoveries" but, for the most part, rather mundane applications of current knowledge, such as better methods for designing drugs, improved tests for genetic disorders, more discerning brain scans, and the like. Some predictions were negative in nature. "Anyone who expects any human-like intelligence from a computer in the next 50 years is doomed to disappointment," proclaimed the physicist and Nobel laureate Philip Anderson.

Second, Koshland's story about the commissioner of patents was apocryphal. In 1940, a scholar named Eber Jeffery examined the patent commissioner anecdote in an article entitled "Nothing Left to Invent," published in the *Journal of the Patent Office Society*.[18] Jeffery traced the story to congressional testimony delivered in 1843 by Henry Ellsworth, then the commissioner of patents. Ellsworth remarked at one point, "The advancement of the arts, from year to year, taxes our credulity and seems to presage the arrival of that period when human improvement must end."

But Ellsworth, far from recommending that his office be shut down, asked for extra funds to cope with the flood of inventions he expected in agriculture, transportation, and communications. Ellsworth did indeed resign two years later, in 1845, but in his resignation letter he made no reference to closing the patent office; rather, he expressed pride at having expanded it. Jeffery concluded that Ellsworth's statement about "that period when human improvement must end" represented "a mere rhetorical flourish intended to emphasize the remarkable strides forward in inventions then current and to be expected in the future." But perhaps Jeffery was not giving Ellsworth enough credit. Ellsworth was, after all, anticipating the argument that Gunther Stent would make more than a century later: the faster science moves, the faster it will reach its ultimate, inevitable limits.

Consider the implications of the alternative position, the one implicitly advanced by Daniel Koshland. He insisted that because science has advanced so rapidly over the past century or so, it can and will continue to do so, possibly forever. But this inductive argument is deeply flawed. Science has only existed for a few hundred years, and its most spectacular achievements have occurred within the last century. Viewed from a historical perspective, the modern era of rapid scientific and technological progress appears to be not a permanent feature of reality, but an aberration, a fluke, a product of a singular convergence of social, intellectual, and political factors.

The Rise and Fall of Progress

In his 1932 book, *The Idea of Progress*, the historian J. B. Bury stated: "Science has been advancing without interruption during the last three or four hundred years; every new discovery has led to new problems and new methods of solution, and opened up new fields for exploration. Hitherto men of science have not been compelled to halt, they have always found means to advance further. But *what assurance have we that they will not come up against impassable barriers?*" [Italics in the original.][19]

Bury had demonstrated through his own scholarship that the concept of progress was only a few hundred years old at most. From the era of the

Roman Empire through the Middle Ages, most truth seekers had a degenerative view of history; the ancient Greeks had achieved the acme of mathematical and scientific knowledge, and civilization had gone downhill from there. Those who followed could only try to recapture some remnant of the wisdom epitomized by Plato and Aristotle. It was such founders of modern, empirical science as Isaac Newton, Francis Bacon, René Descartes, and Gottfried Leibniz who first set forth the idea that humans could systematically acquire and accumulate knowledge through investigations of nature. These ur-scientists believed that the process would be finite, that we could attain complete knowledge of the world and then construct a perfect society, a utopia, based on that knowledge and on Christian precepts. (The new Polynesia!)

Only with the advent of Darwin did certain intellectuals become so enamored with progress that they insisted it might be, or should be, *eternal.* "In the wake of the publication of Darwin's *On the Origin of Species*," Gunther Stent wrote in his 1978 book, *The Paradoxes of Progress*, "the idea of progress was raised to the level of a scientific religion. . . . This optimistic view came to be so widely embraced in the industrialized nations . . . that the claim that progress could presently come to an end is now widely regarded [to be] as outlandish a notion as was in earlier times the claim that the Earth moves around the sun."[20]

It is not surprising that modern nation-states became fervent proponents of the science-is-infinite creed. Science spawned such marvels as The Bomb, nuclear power, jets, radar, computers, and missiles. In 1945 the physicist Vannevar Bush (a distant relative of former president George Bush) proclaimed in *Science: The Endless Frontier* that science was "a largely unexplored hinterland" and an "essential key" to U.S. military and economic security.[21] Bush's essay served as a blueprint for the construction of the National Science Foundation and other federal organizations that thereafter supported basic research on an unparalleled scale.

The Soviet Union was perhaps even more devoted than its capitalist rival to the concept of scientific and technological progress. The Soviets seemed to have taken their lead from Friedrich Engels, who in *Dialectics of Nature* sought to show off his grasp of Newton's inverse square law of gravity in the following passage.

What Luther's burning of the papal Bull was in the religious field, in the field of natural science was the great work of Copernicus. . . . But from then on the development of science went forward in great strides, increasing, so to speak, proportionately to the square of the distance in time of its point of departure, as if it wanted to show the world that for the motion of the highest product of organic matter, the human mind, the law of inverse squares holds good, as it does for the motion of inorganic matter.[22]

Science, in the view of Engels, could and would continue striding forward, at an accelerating pace, forever.

Of course, powerful social, political, and economic forces now oppose this vision of boundless scientific and technological progress. The cold war, which was a major impetus for basic research in the United States and the Soviet Union, is over; the United States and the former Soviet republics have much less incentive to build space stations and gigantic accelerators simply to demonstrate their power. Society is also increasingly sensitive to the adverse consequences of science and technology, such as pollution, nuclear contamination, and weapons of mass destruction.

Even political leaders, who have traditionally been the staunchest defenders of the value of scientific progress, have begun voicing antiscience sentiments. The Czech poet and president Václav Havel declared in 1992 that the Soviet Union epitomized and therefore eternally discredited the "cult of objectivity" brought about by science. Havel expressed the hope that the dissolution of the communist state would bring about "the end of the modern era," which had been "dominated by the culminating belief, expressed in different forms, that the world—and Being as such—is a wholly knowable system governed by a finite number of universal laws that man can grasp and rationally direct for his own benefit."[23]

This disillusionment with science was foreseen early in this century by Oswald Spengler, a German schoolteacher who became the first great prophet of the end of science. In his massive tome, *The Decline of the West*, published in 1918, Spengler argued that science proceeds in a cyclic fashion, with romantic periods of investigation of nature and the invention of new theories giving way to periods of consolidation during which scientific knowledge ossifies. As scientists become more arrogant and less tolerant of other belief systems, notably religious ones, Spengler declared, society will rebel against science and embrace religious fundamentalism and other irrational systems of belief. Spengler predicted that the decline of science and the resurgence of irrationality would begin at the end of this millennium.[24]

Spengler's analysis was, if anything, too optimistic. His view of science as cyclic implied that science might one day be resurrected and undergo a new period of discovery. Science is not cyclic, however, but linear; we can only discover the periodic table and the expansion of the universe and the structure of DNA once. The biggest obstacle to the resurrection of science—and especially pure science, the quest for knowledge about who we are and where we came from—is science's past success.

No More Endless Horizons

Scientists are understandably loath to state publicly that they have entered an era of diminishing returns. No one wants to be recalled as the equivalent of those allegedly shortsighted physicists of a century ago. There is always the danger, too, that predictions of the demise of science will become self-fulfilling. But Gunther Stent is hardly the only prominent scientist to violate the taboo against such prophecies. In 1971, *Science* published an essay titled "Science: Endless Horizons or Golden Age?" by Bentley Glass, an eminent biologist and the president of the American Association for the Advancement of Science, which publishes *Science*. Glass weighed the two scenarios for science's future posited by Vannevar Bush and Gunther Stent and reluctantly came down on the side of Stent. Not only is science finite, Glass argued, but the end is in sight. "We are like the explorers of a great continent," Glass proclaimed, "who have penetrated to its margins in most points of the compass and have mapped the major mountain chains and rivers. There are still innumerable details to fill in, but the endless horizons no longer exist."[25]

According to Glass, a close reading of Bush's *Endless Frontier* essay suggests that he, too, viewed science as a finite enterprise. Nowhere did Bush specifically state that any fields of science could continue generating new discoveries forever. In fact, Bush described scientific knowledge as an "edifice" whose form "is predestined by the laws of logic and the nature of human reasoning. It is almost as though it already existed." Bush's choice of this metaphor, Glass commented, reveals that he considered scientific knowledge to be finite in extent. Glass proposed that the "bold title" of Bush's essay was "never intended to be taken literally, but supposed merely to imply that from our present viewpoint so much yet remains before us to be discovered that the horizons seem virtually endless."

In 1979, in the *Quarterly Review of Biology*, Glass presented evidence to back up his view that science was approaching a culmination.[26] His analysis of the rate of discoveries in biology showed that they had not kept pace with the exponential increase in researchers and funding. "We have been so impressed by the undeniable acceleration in the rate of magnificent achievements that we have scarcely noticed that we are well into an era of diminishing returns," Glass stated. "That is, more and more scientific effort and expenditure of money must be allocated in order to sustain our progress. Sooner or later this will have to stop, because of the insuperable limits to scientific manpower and expenditure. So rapid has been the growth of science in our own century that we have been deluded into thinking that such a rate of progress can be maintained indefinitely."

When I spoke to him in 1994, Glass confessed that many of his colleagues had been dismayed that he had even broached the subject of science's limits, let alone prophesied its demise.[27] But Glass had felt then, and still felt, that the topic was too important to ignore. Obviously science as a social enterprise has *some* limits, Glass said. If science had continued to grow at the same rate as it did earlier in this century, he pointed out, it would soon have consumed the entire budget of the industrialized world. "I think it's rather evident to everybody that there must be brakes put on the amount of funding for science, pure science." This slowdown, he observed, was evident in the decision of the U.S. Congress in 1993 to terminate the superconducting supercollider, the gargantuan particle accelerator that physicists had hoped would propel them beyond quarks and electrons into a deeper realm of microspace, all for a mere $8 billion.

Even if society were to devote all its resources to research, Glass added, science would one day still reach the point of diminishing returns. Why? Because science *works*; it solves its problems. After all, astronomers have already plumbed the farthest reaches of the universe; they cannot see what, if anything, lies beyond its borders. Moreover, most physicists think that the reduction of matter into smaller and smaller particles will eventually end, or may have already ended for all practical purposes. Even if physicists unearth particles buried beneath quarks and electrons, that knowledge will make little or no difference to biologists, who have learned that the most significant biological processes occur at the molecular level and above. "There's a limit to biology there," Glass explained, "that you don't expect to be able to ever break through, just because of the nature of the constitution of matter and energy."

In biology, Glass said, the great revolutions may be in the past. "It's hard to believe, for me, anyway, that anything as comprehensive and earthshaking as Darwin's view of the evolution of life or Mendel's understanding of the nature of heredity will be easy to come by again. After all, these have been discovered!" Biologists certainly have much to learn, Glass emphasized, about diseases such as cancer and AIDS; about the process whereby a single fertilized cell becomes a complex, multicellular organism; about the relation between brain and mind. "There are going to be new additions to the structure of knowledge. But we have made some of the biggest possible advances. And it's just a question of whether there are any more really big changes in our conceptual universe that are going to be made."

Hard Times Ahead for Physics

In 1992, the monthly journal *Physics Today* published an essay entitled "Hard Times," in which Leo Kadanoff, a prominent physicist at the University of Chicago, painted a bleak picture for the future of physics. "Nothing we do is likely to arrest our decline in numbers, support or social value," Kadanoff declared. "Too much of our base depended on events that are now becoming ancient history: nuclear weapons and radar during World War II, silicon and laser technology thereafter, American optimism and industrial hegemony, socialist belief in rationality as a way of improving the world." Those conditions have largely vanished, Kadanoff contended; both physics and science as a whole are now besieged by environmentalists, animal-rights activists, and others with an antiscientific outlook. "In recent decades, science has had high rewards and has been at the center of social interest and concern. We should not be surprised if this anomaly disappears."[28]

Kadanoff, when I spoke to him over the telephone two years later, sounded even gloomier than he had been in his essay.[29] He laid out his worldview for me with a muffled melancholy, as if he were suffering from an existential head cold. Rather than discussing science's social and political problems, as he had in his article in *Physics Today*, he focused on another obstacle to scientific progress: science's past achievements. The great task of modern science, Kadanoff explained, has been to show that the world conforms to certain basic physical laws. "That is an issue which has been explored at least since the Renaissance and maybe a much longer period of time. For me, that's a settled issue. That is, it seems to me that the world *is* explainable by law." The most fundamental laws of nature are embodied in the theory of general relativity and in the so-called standard model of particle physics, which describes the behavior of the quantum realm with exquisite precision.

Just a half century ago, Kadanoff recalled, many reputable scientists still clung to the romantic doctrine of vitalism, which holds that life springs from some mysterious élan vital that cannot be explained in terms of physical laws. As a result of the findings of molecular biology—beginning with the discovery of the structure of DNA in 1953—"there are relatively few well-educated people" who admit to belief in vitalism, Kadanoff said.

Of course, scientists still have much to learn about how the fundamental laws generate "the richness of the world as we see it." Kadanoff himself is a leader in the field of condensed-matter physics, which studies the behavior not of individual subatomic particles, but of solids or liquids. Kadanoff has also been associated with the field of chaos, which addresses phenomena that unfold in predictably unpredictable ways. Some proponents of chaos—and of a closely related field called complexity—have suggested that with the help of powerful computers and new mathematical methods they will discover truths that surpass those revealed by the "reductionist" science of the past. Kadanoff has his doubts. Studying the consequences of fundamental laws is "in a way less interesting" and "less deep," he said, than showing that the world is lawful. "But now that we know the world is lawful," he added, "we have to go on to other things. And yes, it probably excites the imagination of the average human being less. Maybe with good reason."

Kadanoff pointed out that particle physics has not been terribly exciting lately either. Experiments over the past few decades have merely confirmed existing theories rather than revealing new phenomena requiring new laws; the goal of finding a unified theory of all of nature's forces seems impossibly distant. In fact, no field of science has yielded any truly deep discoveries for a long time, Kadanoff said. "The truth is, there is nothing—there is *nothing*—of the same order of magnitude as the accomplishments of the invention of quantum mechanics or of the double helix or of relativity. Just nothing like that has happened in the last few decades." Is this state of affairs permanent? I asked. Kadanoff was silent for a moment. Then he sighed, as if trying to exhale all his world-weariness. "Once you have proven that the world is lawful," he replied, "to the satisfaction of many human beings, you can't do that again."

Whistling to Keep Our Courage Up

One of the few modern philosophers to devote serious thought to the limits of science is Nicholas Rescher of the University of Pittsburgh. In his 1978 book, *Scientific Progress*, Rescher deplored the fact that Stent, Glass, and other prominent scientists seemed to think that science might be approaching a cul-de-sac. Rescher intended to provide "an antidote to this currently pervasive tendency of thought" by demonstrating that science is at least potentially infinite.[30] But the scenario he sketched out over the course of his book was hardly optimistic. He argued that science, as a fundamentally empirical, experimental discipline, faces economic constraints. As scientists try to extend their theories into more remote domains—seeing further into the universe, deeper into matter—their costs will inevitably escalate and their returns diminish.

"Scientific innovation is going to become more and more difficult as we push out further and further from our home base toward more remote frontiers. If the present perspective is even partly correct, the half-millennium commencing around 1650 will eventually come to be regarded among the great characteristic developmental transformations of human history, with the age of The Science Explosion as unique in its own historical structure as The Bronze Age or The Industrial Revolution or The Population Explosion."[31]

Rescher tacked what he apparently thought was a happy coda onto his depressing scenario: science will never end; it will just go slower and slower and slower, like Zeno's tortoise. Nor should scientists ever conclude that their research must degenerate into the mere filling in of details; it is always *possible* that one of their increasingly expensive experiments will have revolutionary import, comparable to that of quantum mechanics or Darwinian theory.

Bentley Glass, in a review of Rescher's book, called these prescriptions "whistling to keep one's courage up in the face of what, for most practitioners of science, is a bleak and imminent prospect."[32] When I telephoned Rescher in August 1992, he acknowledged that his analysis had been in most respects a grim one. "We can only investigate nature by interacting with it," he said. "To do that we must push into regions never investigated before, regions of higher density, lower temperature, or higher energy. In all these cases we are pushing fundamental limits, and that requires ever more elaborate and expensive apparatuses. So there is a limit imposed on science by the limits of human resources."

But Rescher insisted that "big plums, first-rate discoveries," might—must!—lie ahead. He could not say whence those discoveries might arise. "It's like the jazz musician who was asked where jazz is going, and he said, 'If I knew we'd be there by now.'" Rescher, finally, fell back on the that's-what-they-thought-at-the-end-of-the-last-century argument. The fact that such scientists as Stent, Glass, and Kadanoff seemed to fear that science was drawing to a close, Rescher said, gave him confidence that some marvelous discovery was pending. Rescher, like many other would-be seers, had succumbed to wishful thinking. He admitted that he felt that the end of science would be a tragedy for humanity. If the quest for knowledge ended, what would become of us? What would give our existence meaning?

The Meaning of Francis Bacon's Plus Ultra

The second most common response to the suggestion that science is ending—after "that's what they thought at the end of the last century"—is the old maxim "Answers raise new questions." Kant wrote in *Prolegomena to Any Future Metaphysics* that "every answer given on principles of experience begets a fresh question, which likewise requires its answer and thereby clearly shows the insufficiency of all physical modes of explanation to satisfy reason."[33] But Kant also suggested (anticipating the arguments of Gunther Stent), that the innate structure of our minds constrains both the questions we put to nature and the answers we glean from it.

Of course, science will continue to raise new questions. Most are trivial, in that they concern details that do not affect our basic understanding of nature. Who really cares, except specialists, about the precise mass of the top quark, the existence of which was finally confirmed in 1994 after research costing billions of dollars? Other questions are profound but unanswerable. In fact, the most persistent foil to the *completion* of science—to the attainment of the truly satisfying theory that Roger Penrose and others dream of—is the human ability to invent unanswerable questions. Presented with a purported theory of everything, someone always can and will ask: But how do we know that quarks or even superstrings (in the unlikely event that they are one day shown to exist) are not composed of still smaller entities—ad infinitum? How do we know that the visible universe is not just one of an infinite number of universes? Was our universe necessary or just a cosmic fluke? What about life? Are computers capable of conscious thought? Are amoebas?

No matter how far empirical science goes, our imaginations can always go farther. That is the greatest obstacle to the hopes—and fears—of scientists that we will find *The Answer*, a theory that quenches our curiosity forever. Francis Bacon, one of the founders of modern science, expressed his belief in the vast potential of science with the Latin term *plus ultra*, "more beyond."[34] But *plus ultra* does not apply to science per se, which is a tightly constrained method for examining nature. *Plus ultra* applies, rather, to our imaginations. Although our imaginations are constrained by our evolutionary history, they will always be capable of venturing beyond what we truly know.

Even in the new Polynesia, Gunther Stent suggested, a few persistent souls will keep striving to transcend the received wisdom. Stent called these truth seekers "Faustian" (a term he borrowed from Oswald Spengler). I call them strong scientists (a term I coopted from Harold Bloom's *The Anxiety of Influence*). By raising questions that science cannot answer, strong scientists can continue the quest for knowledge in the speculative mode that I call ironic science even after empirical science—the kind of science that answers questions—has ended.

The poet John Keats coined the term *negative capability* to describe the ability of certain great poets to remain "in uncertainties, mysteries, doubts, without any irritable reaching after fact and reason." As an example, Keats singled out his fellow poet Samuel Coleridge, who "would let go by a fine isolated verisimilitude caught from the penetralium of mystery, from being incapable of remaining content with half-knowledge."[35] The most important function of ironic science is to serve as humanity's negative capability. Ironic science, by raising unanswerable questions, reminds us that all our knowledge is half-knowledge; it reminds us of how little we know. But ironic science does not make any significant contributions to knowledge itself. Ironic science is thus less akin to science in the traditional sense than to literary criticism—or to philosophy.

CHAPTER TWO

The End of Philosophy

Twentieth-century science has given rise to a marvelous paradox. The same extraordinary progress that has led to predictions that we may soon know everything that can be known has also nurtured doubts that we can know *anything* for certain. When one theory so rapidly succeeds another, how can we ever be sure that any theory is true? In 1987 two British physicists, T. Theocharis and M. Psimopoulos, excoriated this skeptical philosophical position in an essay entitled "Where Science Has Gone Wrong." Published in the British journal *Nature*, the essay blamed the "deep and widespread malaise" in science on philosophers who had attacked the notion that science could achieve objective knowledge. The article printed photographs of four particularly egregious "betrayers of the truth": Karl Popper, Imre Lakatos, Thomas Kuhn, and Paul Feyerabend.[1]

The photographs were grainy, black-and-white shots of the sort that adorn a lurid exposé about a venerable banker who has been caught swindling retirees. These, clearly, were intellectual transgressors of the worst sort. Feyerabend, whom the essayists called "the worst enemy of science," was the most wicked-looking of the bunch. Smirking at the camera over glasses perched on the tip of his nose, he was clearly either anticipating or relishing the perpetration of some diabolical prank. He looked like an intellectual version of Loki, the Norse god of mischief.

The main complaint of Theocharis and Psimopoulos was silly. The skepticism of a few academic philosophers has never represented a serious threat to the massive, well-funded bureaucracy of science. Many scientists, particularly would-be revolutionaries, find the ideas of Popper et al. comforting; if our current knowledge is provisional, there is always the possibility that great revelations lie ahead. Theocharis and Psimopoulos did make one intriguing assertion, however, that the ideas of the skeptics are "flagrantly self-refuting—they negate and destroy themselves." It would be interesting, I thought, to put this argument to the philosophers and see how they responded.

Eventually I had the opportunity to do just that with all the "betrayers of the truth" except for Lakatos, who died in 1974. During my interviews, I also tried to find out whether these philosophers were really as skeptical, as doubtful of science's ability to achieve truth, as some of their own statements implied. I came away convinced that Popper, Kuhn, and Feyerabend each believed very much in science; in fact, their skepticism was motivated by their belief. Their biggest failing, perhaps, was to credit science with more power than it actually has. They feared that science might extinguish our sense of wonder and therefore bring science itself—and all forms of knowledge seeking—to an end. They were trying to protect humanity, scientists included, from the naive faith in science exemplified by such scientists as Theocharis and Psimopoulos.

As science has grown in power and prestige over the past century, too many philosophers have served as science's public relations agents. This trend can be traced to such thinkers as Charles Sanders Peirce, an American who founded the philosophy of pragmatism but could not keep a job or a wife and died penniless and miserable in 1914. Peirce offered this definition of absolute truth: it is whatever scientists say it is when they come to the end of their labors.[2]

Much of philosophy since Peirce has merely elaborated on his view. The dominant philosophy in Europe early in this century was logical positivism, which asserted that we can only know that something is true if it can be logically or empirically demonstrated. The positivists upheld mathematics and science as the supreme sources of truth. Popper, Kuhn, and Feyerabend—each in his own way and for his own reasons—sought to counter this fawning attitude toward science. These philosophers realized that in an age when science is ascendant, the highest calling of philosophy should be to serve as the negative capability of science, to infuse scientists with doubt. Only thus can the human quest for knowledge remain open-ended, potentially infinite; only thus can we remain awestruck before the mystery of the cosmos.

Of the three great skeptics I interviewed, Popper was the first to make his mark.[3] His philosophy stemmed from his effort to distinguish pseudoscience, such as Marxism or astrology or Freudian psychology, from genuine science, such as Einstein's theory of relativity. The latter, Popper decided, was testable; it made predictions about the world that could be empirically checked. The logical positivists had said as much. But Popper denied the positivist assertion that scientists can *prove* a theory through induction, or repeated empirical tests or observations. One never knows if one's observations have been sufficient; the next observation might contradict all that preceded it. Observations can never prove a theory but can only disprove, or falsify it. Popper often bragged that he had "killed" logical positivism with this argument.[4]

Popper expanded his falsification tenet into a philosophy that he called critical rationalism. One scientist ventures a proposal and others try to bat it down with contrary arguments or experimental evidence. Popper viewed criticism, and even conflict, as essential for progress of all kinds. Just as scientists approach the truth through what he calls "conjecture and refutation," so do species evolve through competition and societies through political debate. A "human society without conflict," he once wrote, "would be a society not of friends but of ants."[5] In *The Open Society and Its Enemies*, published in 1945, Popper asserted that politics, even more than science, required the free play of ideas and criticism. Dogmatism inevitably led not to utopia, as Marxists and fascists alike claimed, but to totalitarian repression.

I began to discern the paradox lurking at the heart of Popper's work—and persona—when, prior to meeting him, I asked other philosophers about him. Queries of this kind usually elicit rather dull, generic praise, but in this case my interlocutors had nothing good to say. They revealed that this man who inveighed against dogmatism was himself almost pathologically dogmatic and demanding of fealty from students. There was an old joke about Popper: *The Open Society and Its Enemies* should have been titled *The Open Society by One of Its Enemies*.

To arrange my interview with Popper, I telephoned the London School of Economics, where he had taught since the late 1940s. A secretary there said that Popper generally worked at his home in Kensington, an affluent district of London, and gave me his number. I called, and a woman with an imperious, German-accented voice answered. Mrs. Mew, housekeeper and assistant to "Sir Karl." Before Sir Karl would see me, I had to send her a sample of my writings. She gave me a reading list that would prepare me for my meeting: a dozen or so books by Sir Karl. Eventually, after numerous faxes and telephone calls, she set a date. She also provided directions to the train station near Sir Karl's house. When I asked her for directions from the train station to the house, Mrs. Mew assured me that all the cab drivers knew where Sir Karl lived. "He's quite famous."

"Sir Karl Popper's house, please," I said as I climbed into a cab at Kensington station. "Who?" the driver replied. Sir Karl Popper? The famous philosopher? Never heard of him, the driver said. He was familiar with the street on which Popper lived, however, and we found Popper's home—a two-story cottage surrounded by scrupulously trimmed grass and shrubs—with little difficulty.[6]

A tall, handsome woman in black pants and shirt, with short, dark hair brushed straight back, answered the door: Mrs. Mew. She was only slightly less forbidding in person than over the telephone. As she led me into the house, she told me that Sir Karl was quite tired. He had undergone a spate of interviews and congratulations brought on by his 90th birthday the previous month, and he had been working too hard preparing an acceptance speech for the Kyoto Award, known as Japan's Nobel. I should expect to speak to him for only an hour at the most.

I was trying to lower my expectations when Popper made his entrance. He was stooped, equipped with a hearing aid, and surprisingly short; I had assumed that the author of such autocratic prose would be tall. Yet he was as kinetic as a bantamweight boxer. He brandished an article I had written for *Scientific American* about how quantum mechanics was compelling some physicists to abandon the view of physics as a wholly objective enterprise.[7] "I don't believe a word of it," he declared in an Austrian-accented growl. "Subjectivism" has no place in physics, quantum or otherwise. "Physics," he exclaimed, grabbing a book from a table and slamming it down, "is that!" (This from a man who cowrote a book espousing dualism, the notion that ideas and other constructs of the human mind exist independently of the material world.)[8]

Once seated, he kept darting away to forage for books or articles that could buttress a point. Striving to dredge a name or date from his memory, he kneaded his temples and gritted his teeth as if in agony. At one point, when the word *mutation* briefly eluded him, he slapped his forehead repeatedly and with alarming force, shouting, "Terms, terms, terms!"

Words poured from him so rapidly and with so much momentum that I began to lose hope that I could ask any of my prepared questions. "I am over 90 and I can still think," he declared, as if suspecting that I doubted it. He tirelessly touted a theory of the origin of life proposed by a former student, Günther Wächtershäuser, a German patent attorney who had a Ph.D. in chemistry.[9] Popper kept emphasizing that he had known all the titans of twentieth-century science: Einstein, Schrödinger, Heisenberg. Popper blamed Bohr, whom he knew "very well," for having introduced subjectivism into physics. Bohr was "a marvelous physicist, one of the greatest of all time, but he was a miserable philosopher, and one couldn't talk to him. He was talking all the time, allowing practically only one or two words to you and then at once cutting in."

As Mrs. Mew turned to leave, Popper abruptly asked her to find one of his books. She disappeared for a few minutes and then returned empty-handed. "Excuse me, Karl, I couldn't find it," she reported. "Unless I have a description, I can't check every bookcase."

"It was actually, I think, on the right of this corner, but I have taken it away maybe . . ." His voice trailed off. Mrs. Mew somehow rolled her eyes without really rolling them and vanished.

He paused a moment, and I desperately seized the opportunity to ask a question. "I wanted to ask you about . . ."

"Yes! You should ask me your questions! I have wrongly taken the lead. You can ask me all your questions first."

As I began to question Popper about his views, it became apparent that his skeptical philosophy stemmed from a deeply romantic, idealized view of science. He thus denied the assertion, often made by the logical positivists, that science can ever be reduced to a formal, logical system, in which raw data are methodically converted into truth. A scientific theory, Popper insisted, is an invention, an act of creation as profoundly mysterious as anything in the arts. "The history of science is everywhere speculative," Popper said. "It is a marvelous history. It makes you proud to be a human being." Framing his face in his outstretched hands, Popper intoned, "I believe in the human mind."

For similar reasons, Popper had battled throughout his career against the doctrine of scientific determinism, which he felt was antithetical to human creativity and freedom and thus to science itself. Popper claimed to have realized long before modern chaos theorists that not only quantum systems but even classical, Newtonian ones are inherently unpredictable; he had delivered a lecture on this subject in the 1950s. Waving at the lawn outside the window he said, "There is chaos in every grass."

When I asked Popper if he thought that science was incapable of achieving absolute truth, he exclaimed, "No no!" and shook his head vehemently. He, like the logical positivists before him, believed that a scientific theory could be "absolutely" true. In fact, he had "no doubt" that some current scientific theories were absolutely true (although he refused to say which ones). But he rejected the positivist belief that we can ever *know* that a theory is true. "We must distinguish between truth, which is objective and absolute, and certainty, which is subjective."

If scientists believed too much in their own theories, Popper realized, they might stop seeking truth. And that would be a tragedy, since for Popper truth seeking was what made life worth living. "To search for the truth is a kind of religion," he said, "and I think it is also an ethical belief." Popper's conviction that the search for knowledge must never cease is reflected in the title of his autobiography, *Unended Quest*.

He thus scoffed at the hope of some scientists to achieve a complete theory of nature, one that answers all questions. "Many people think that the problems can be solved; many people think the opposite. I think we have gone very far, but we are much further away. I must show you one passage that bears on this." He shuffled off again and returned with his book *Conjectures and Refutations*. Opening it, he read his own words with reverence: "In our infinite ignorance we are all equal."

Popper also believed that science could never answer questions about the meaning and purpose of the universe. For these reasons he had never completely repudiated religion, although he had long ago abandoned the Lutheranism of his youth. "We know very little, and we should be modest and not pretend we know anything about ultimate questions of this kind."

Yet Popper abhorred those modern philosophers and sociologists who claim that science is incapable of achieving *any* truth and who argue that scientists adhere to theories for cultural and political reasons rather than rational ones. Such critics, Popper charged, resent being viewed as inferior to genuine scientists and are trying to "change their status in the pecking order." I suggested that these critics sought to describe how science *is* practiced, whereas he, Popper, tried to show how it *should* be practiced. Somewhat to my surprise, Popper nodded. "That is a very good statement," he said. "You can't see what science is without having in your head an idea of what science should be." Popper had to agree that scientists often fell short of the ideal he had set for them. "Since scientists got subsidies for their work, science isn't exactly what it should be. This is unavoidable. There is a certain corruption, unfortunately. But I don't talk about that."

Popper then proceeded to talk about it. "Scientists are not as self-critical as they should be," he asserted. "There is a certain wish that you, people like you"—he jabbed a finger at me—"should bring them before the public." He stared at me a moment, then reminded me that he had not sought this interview. "Far from it," he said. "You know that I have really made not only no attempt but have not encouraged you." Popper then plunged into an excruciatingly technical critique—involving triangulation and other arcana—of the big bang theory. "It's always the same," he summed up. "The difficulties are underrated. It is presented in a spirit as if this all has scientific certainty, but scientific certainty doesn't exist."

I asked Popper if he felt that biologists were also too committed to Darwin's theory of natural selection; in the past he had suggested that the theory was tautological and thus pseudoscientific.[10] "That was perhaps going too far," Popper said, waving his hand dismissively. "I'm not dogmatic about my own views." Suddenly, he pounded the table and exclaimed, "One ought to look for alternative theories! This"—he waved the paper by Günther Wächtershäuser on the origin of life—"is an alternative theory. It seems to be a better theory." That doesn't mean that the theory is true, Popper quickly added. "The origin of life will forever remain untestable, probably," Popper said. Even if scientists create life in a laboratory, he explained, they can never be sure that life actually began in the same way.

It was time to launch my big question. Was his own falsification concept falsifiable? Popper glared at me. Then his expression softened, and he placed his hand on mine. "I don't want to hurt you," he said gently, "but it is a silly question." Peering searchingly into my eyes, he inquired whether one of his critics had persuaded me to ask this question. Yes, I lied. "Exactly," he said, looking pleased.

"The first thing you do in a philosophy seminar when somebody proposes an idea is to say it doesn't satisfy its own criteria. It is one of the most idiotic criticisms one can imagine!" His falsification concept, he said, is a criterion for distinguishing between empirical modes of knowledge, namely, science, and nonempirical ones, such as philosophy. Falsification itself is "decidably unempirical"; it belongs not to science but to philosophy, or "metascience," and it does not even apply to all of science. Popper was admitting, essentially, that his critics were right: falsification is a mere guideline, a rule of thumb, sometimes helpful and sometimes not.

Popper said he had never before responded to the question I had just posed. "I found it too stupid to be answered. You see the difference?" he asked, his voice gentle again. I nodded. The question seemed a bit silly to me, too, I said, but I just thought I should ask. He smiled and squeezed my hand, murmuring, "Yes, very good."

Since Popper seemed so agreeable, I mentioned that one of his former students had accused him of not tolerating criticism of his own ideas. Popper's eyes blazed. "It is completely untrue! I was *happy* when I got criticism! Of course, not when I would answer the criticism, like I have answered it when you gave it to me, and the person would still go on with it. That is the thing which I found uninteresting and would not tolerate." If that happened, Popper would order the student out of his class.

The light in the kitchen was acquiring a ruddy hue when Mrs. Mew stuck her head in the door and informed us that we had been talking for more than three hours. How much longer, she inquired a bit peevishly, did we expect to continue? Perhaps she had better call me a cab? I looked at Popper, who had broken into a bad-boy grin but did appear to be drooping.

I slipped in a final question: Why in his autobiography did Popper say that he was the happiest philosopher he knew? "Most philosophers are really deeply depressed," he replied, "because they can't produce anything worthwhile." Looking pleased with himself, Popper glanced over at Mrs. Mew, who wore an expression of horror. Popper's own smile abruptly faded. "It would be better not to write that," he said, turning back to me. "I have enough enemies, and I better not answer them in this way." He stewed a moment and added, "But it is so."

I asked Mrs. Mew if I could have a copy of the speech Popper was going to deliver at the Kyoto Award ceremony in Japan. "No, not now," she said curtly. "Why not?" Popper inquired. "Karl," she replied, "I've been typing the second lecture nonstop, and I'm a bit . . ." She sighed. "You know what I mean?" Anyway, she added, she did not have a final version. "What about an uncorrected version?" Popper asked. Mrs. Mew stalked off.

She returned and shoved a copy of Popper's lecture at me. "Have you got a copy of *Propensities*?" Popper asked her.[11] She pursed her lips and stomped into the room next door, while Popper explained the book's theme to me. The lesson of quantum mechanics and even of classical physics, Popper said, is that nothing is determined, nothing is certain, nothing is completely predictable; there are only propensities for certain things to occur. "For example," Popper added, "in this moment there is a certain propensity that Mrs. Mew may find a copy of my book."

"Oh, please!" Mrs. Mew exclaimed from the next room. She returned, no longer making any attempt to hide her annoyance. "Sir Karl, Karl, you have given away the last copy of *Propensities*. Why do you do that?"

"The last copy was given away in your presence," he declared.

"I don't think so," she retorted. "Who was it?"

"I can't remember," he muttered sheepishly.

Outside, a black cab pulled into the driveway. I thanked Popper and Mrs. Mew for their hospitality and took my leave. As the cab pulled away, I asked the driver if he knew whose house this was. No, he didn't. Someone famous, was it? Yes, actually: Sir Karl Popper. Who? Karl Popper, I replied, one of the greatest philosophers of the twentieth century. "Is that right?" murmured the driver.

Popper has always been popular among scientists—and with good reason, since he depicted science as an endless romantic adventure. An editorial in *Nature* once called Popper, quite justly, "the philosopher *for* science" [italics added].[12] But Popper's fellow philosophers have been less kind. His oeuvre, they point out, is rife with contradictions. Popper argued that science could not be reduced to a method, but his falsification scheme was just such a method. Moreover, the arguments that he used to kill the possibility of absolute verification could also be used to kill falsification. If it is always possible that future observations will contradict a theory, then it is also possible that future observations may resurrect a theory that has previously been falsified. It is more reasonable to assume, critics of Popper have argued, that just as some scientific theories can be falsified, so some can be confirmed; there is no point in remaining uncertain, after all, that the earth is round and not flat.

When Popper died in 1994, two years after I met him, the *Economist* hailed him as having been "the best-known and most widely read of living philosophers."[13] It praised, in particular, his insistence on antidogmatism in the political realm. But the obituary also noted that Popper's treatment of induction (the basis of his falsification scheme) had been rejected by later philosophers. "According to his own theories, Popper should have welcomed this fact," the *Economist* noted drily, "but he could not bring himself to do so. The irony is that, here, Popper could not admit he was wrong." Popper's antidogmatism, when applied to science, had become a kind of dogmatism.

Although Popper abhorred psychoanalysis, his own work, finally, may be best understood in psychoanalytic terms. His relationship with authority figures—from scientific giants, such as Bohr, to his assistant, Mrs. Mew—was obviously complex, alternating between defiance and deference. In what is perhaps the single most revealing passage in his autobiography, Popper mentioned that his parents were both Austrian Jews who had converted to Lutheranism. He then argued that the failure of other Jews to assimilate themselves into Germanic culture—and their prominent roles in leftist politics—contributed to the emergence of fascism and state-sponsored anti-Semitism in the 1930s: ". . . anti-Semitism was an evil, to be feared by Jews and non-Jews alike, and . . . it was the task of all people of Jewish origin to do their best not to provoke it."[14] Popper was coming close to blaming the Jews for the Holocaust.

The Structure of Thomas Kuhn

"Look," Thomas Kuhn said. The word was weighted with weariness, as if Kuhn was resigned to the fact that I would misinterpret him, but he was still going to try—no doubt in vain—to make his point. Kuhn uttered the word often. "Look," he said again. He leaned his gangly frame and long face forward, and his big lower lip, which ordinarily curled up amiably at the corners, sagged. "For Christ's sake, if I had my choice of having written the book or not having written it, I would choose to have written it. But there have certainly been aspects involving considerable upset about the response to it."

"The book" was *The Structure of Scientific Revolutions*, which may be the most influential treatise ever written on how science does (or does not) proceed. It is notable for having spawned the trendy term *paradigm*. It also fomented the now trite idea that personalities and politics play a large role in science. The book's most profound argument was less obvious: scientists can never truly understand the real world or even each other.[15]

Given this theme, one might think that Kuhn would have expected his own message to be at least partially misunderstood. But when I interviewed Kuhn in his office at the Massachusetts Institute of Technology (of all places) some three decades after the publication of *Structure*, he seemed to be deeply pained by the breadth of misunderstanding of his book. He was particularly upset by claims that he had described science as irrational. "If they had said '*a*rational' I wouldn't have minded at all," he said with no trace of a smile.

Kuhn's fear of compounding the confusion over his work had made him a bit press shy. When I first telephoned him to ask for an interview, he turned me down. "Look. I think not," he said. He revealed that *Scientific American*, my employer, had given *Structure* "the worst review I can remember." (The squib was indeed dismissive; it called Kuhn's argument "much ado about very little." But what did Kuhn expect from a magazine that celebrates science?)[16] Pointing out that I had not been at the magazine then—the review ran in 1964—I begged him to reconsider. Kuhn finally, reluctantly, agreed.

When we at last sat down together in his office, Kuhn expressed nominal discomfort at the notion of delving into the roots of his thought. "One is not one's own historian, let alone one's own psychoanalyst," he warned me. He nonetheless traced his view of science to an epiphany he experienced in 1947, when he was working toward a doctorate in physics at Harvard. While reading Aristotle's *Physics*, Kuhn had become astonished at how "wrong" it was. How could someone who wrote so brilliantly on so many topics be so misguided when it came to physics?

Kuhn was pondering this mystery, staring out his dormitory window ("I can still see the vines and the shade two-thirds of the way down"), when suddenly Aristotle "made sense." Kuhn realized that Aristotle invested basic concepts with different meanings than did modern physicists. Aristotle used the term *motion*, for example, to refer not just to change in position but to change in general—the reddening of the sun as well as its descent toward the horizon. Aristotle's physics, understood on its own terms, was simply different from, rather than inferior to, Newtonian physics.

Kuhn left physics for philosophy, and he struggled for 15 years to transform his epiphany into the theory set forth in *The Structure of Scientific Revolutions*. The keystone of his model was the concept of a paradigm. *Paradigm*, pre-Kuhn, referred merely to an example that serves an educational purpose; *amo, amas, amat,* for instance, is a paradigm for teaching conjugations in Latin. Kuhn used the term to refer to a collection of procedures or ideas that instruct scientists, *implicitly*, what to believe and how to work. Most scientists never question the paradigm. They solve puzzles, problems whose solutions reinforce and extend the scope of the paradigm rather than challenge it. Kuhn called this "mopping up," or "normal science." There are always anomalies, phenomena that the paradigm cannot account for or that even contradict it. Anomalies are often ignored, but if they accumulate they may trigger a revolution (also called a paradigm shift, although not originally by Kuhn), in which scientists abandon the old paradigm for a new one.

Denying the view of science as a continual building process, Kuhn held that a revolution is a destructive as well as a creative act. The proposer of a new paradigm stands on the shoulders of giants (to borrow Newton's phrase) and then bashes them over the head. He or she is often young or new to the field, that is, not fully indoctrinated. Most scientists yield to a new paradigm reluctantly. They often do not understand it, and they have no objective rules by which to judge it. Different paradigms have no common standard for comparison; they are "incommensurable," to use Kuhn's term. Proponents of different paradigms can argue forever without resolving their differences because they invest basic terms—motion, particle, space, time—with different meanings. The conversion of scientists is thus both a subjective and a political process. It may involve sudden, intuitive understanding—like that finally achieved by Kuhn as he pondered Aristotle. Yet scientists often adopt a paradigm simply because it is backed by others with strong reputations or by a majority of the community.

Kuhn's view diverged from Popper's in several important respects. Kuhn (like other critics of Popper) argued that falsification is no more possible than verification; each process implies the existence of absolute standards of evidence, which transcend any individual paradigm. A new paradigm may solve puzzles better than the old one does, and it may yield more practical applications. "But you cannot simply describe the other science as false," Kuhn said. Just because modern physics has spawned computers, nuclear power, and CD players does not mean it is truer, in an absolute sense, than Aristotle's physics. Similarly, Kuhn denied that science is constantly approaching the truth. At the end of *Structure* he asserted that science, like life on earth, does not evolve *toward* anything, but only *away* from something.

Kuhn described himself to me as a "post-Darwinian Kantian." Kant, too, believed that without some sort of *a priori* paradigm the mind cannot impose order on sensory experience. But whereas Kant and Darwin each thought that we are all born with more or less the same innate paradigm, Kuhn argued that our paradigms keep changing as our culture changes. "Different groups, and the same group at different times," Kuhn told me, "can have different experiences and therefore in some sense live in different worlds." Obviously all humans share some responses to experience, simply because of their shared biological heritage, Kuhn added. But whatever is universal in human experience, whatever transcends culture and history, is also "ineffable," beyond the reach of language. Language, Kuhn said, "is not a universal tool. It's not the case that you can say anything in one language that you can say in another."

But isn't mathematics a kind of universal language? I asked. Not really, Kuhn replied, since it has no meaning; it consists of syntactical rules without any semantic content. "There are perfectly good reasons why mathematics can be considered a language, but there is a very good reason why it isn't." I objected that although Kuhn's view of the limits of language might apply to certain fields with a metaphysical cast, such as quantum mechanics, it did not hold in all cases. For example, the claim of a few biologists that AIDS is not caused by the so-called AIDS virus is either right or wrong; language is not the crucial issue. Kuhn shook his head. "Whenever you get two people interpreting the same data in different ways," he said, "that's metaphysics."

So, were his own ideas true or not? "Look," Kuhn responded with even more weariness than usual; obviously he had heard this question many times before. "I think this way of talking and thinking that I am engaged in opens up a range of possibilities that can be investigated. But it, like any scientific construct, has to be evaluated simply for its utility—for what you can do with it."

But then Kuhn, having set forth his bleak view of the limits of science and indeed of all human discourse, proceeded to complain about the many ways in which his book had been misinterpreted and misused, especially by admirers. "I've often said I'm much fonder of my critics than my fans." He recalled students approaching him to say, "Oh, thank you, Mr. Kuhn, for telling us about paradigms. Now that we know about them we can get rid of them." He insisted that he did not believe that science was *entirely* political, a reflection of the prevailing power structure. "In retrospect, I begin to see why this book fed into that, but boy, was it not meant to, and boy, does it not mean to."

His protests were to no avail. He had a painful memory of sitting in on a seminar and trying to explain that the concepts of truth and falsity are perfectly valid, and even necessary—within a paradigm. "The professor finally looked at me and said, 'Look, you don't know how radical this book is.' " Kuhn was also upset to find that he had become the patron saint of all would-be scientific revolutionaries. "I get a lot of letters saying, 'I've just read your book, and it's transformed my life. I'm trying to start a revolution. Please help me,' and accompanied by a book-length manuscript."

Kuhn declared that, although his book was not intended to be pro-science, he *is* pro-science. It is the rigidity and discipline of science, Kuhn said, that makes it so effective at problem solving. Moreover, science produces "the greatest and most original bursts of creativity" of any human enterprise. Kuhn conceded that he was partly to blame for some of the antiscience interpretations of his model. After all, in *Structure* he did call scientists committed to a paradigm "addicts"; he also compared them to the brainwashed characters in Orwell's *1984*.[17] Kuhn insisted that he did not mean to be condescending by using such terms as *mopping up* or *puzzle solving* to describe what most scientists do. "It was meant to be descriptive." He ruminated a bit. "Maybe I should have said more about the glories that result from puzzle solving, but I thought I was doing that."

As for the word *paradigm*, Kuhn conceded that it had become "hopelessly overused" and was "out of control." Like a virus, the word spread beyond the history and philosophy of science and infected the intellectual community at large, where it came to signify virtually any dominant idea. A 1974 *New Yorker* cartoon captured the phenomenon. "Dynamite, Mr. Gerston!" gushed a woman to a smug-looking man. "You're the first person I ever heard use 'paradigm' in real life." The low point came during the Bush administration, when White House officials introduced an economic plan called "the New Paradigm" (which was really just warmed-over Reaganomics).[18]

Kuhn admitted, again, that the fault was partly his, since in *Structure* he had not defined paradigm as crisply as he might have. At one point paradigm referred to an archetypal experiment, such as Galileo's legendary (and probably apocryphal) dropping of weights from the Leaning Tower of Pisa. Elsewhere the term referred to "the entire constellation of beliefs" that binds a scientific community together. (Kuhn denied, however, that he had defined paradigm in 21 different ways, as one critic claimed.)[19] In a postscript to later editions of *Structure*, Kuhn recommended that *paradigm* be replaced with the term *exemplar*, but it never caught on. He eventually gave up all hope of explaining what he really meant. "If you've got a bear by the tail, there comes a point at which you've got to let it go and stand back," he sighed.

One of the sources of *Structure*'s power and persistence is its profound ambiguity; it appeals to relativists and to science worshipers alike. Kuhn acknowledged that "a lot of the success of the book and some of the criticisms are due to its vagueness." (One wonders whether Kuhn's writing style is intended or innate; his speech is as profoundly tangled, as suffused with subjunctives and qualifiers, as is his prose.) *Structure* is clearly a work of literature, and as such it is subject to many interpretations. According to literary theory, Kuhn himself cannot be trusted to provide a definitive account of his own work. Here is one possible interpretation of Kuhn's text, and of Kuhn. Kuhn focused on what science is rather than on what it should be; he had a much more realistic, hard-nosed, psychologically accurate view of science than did Popper. Kuhn realized that, given the power of modern science and the tendency of scientists to believe in theories that have withstood many tests, science may well enter a phase of permanent normalcy, in which no further revolutions, or revelations, are possible.

Kuhn also accepted, as Popper could not, that science might not continue forever, even in a normal state. "There was a beginning to it," Kuhn said. "There are lots of societies that don't have it. It takes very special conditions to support it. Those social conditions are now getting harder to find. Of *course* it could end." Science might even end, Kuhn said, because scientists cannot make any further headway, even given adequate resources.

Kuhn's recognition that science might cease—leaving us with what Charles Sanders Peirce defined as the truth about nature—made it even more imperative for Kuhn than for Popper to challenge science's authority, to deny that science could *ever* arrive at absolute truth. "The one thing I think you shouldn't say is that now we've found out what the world is really like," Kuhn said. "Because that's not what I think the game is about."

Kuhn has tried, throughout his career, to remain true to that original epiphany he experienced in his dormitory at Harvard. During that moment Kuhn saw—he knew!—that reality is ultimately unknowable; any attempt to describe it obscures as much as it illuminates. But Kuhn's insight forced him to take the untenable position that because all scientific theories fall short of absolute, mystical truth, they are all equally untrue; because we cannot discover *The Answer*, we cannot find any answers. His mysticism led him toward a position as absurd as that of the literary sophists who argue that all texts—from *The Tempest* to an ad for a new brand of vodka—are equally meaningless, or meaningful.

At the end of *Structure*, Kuhn briefly raised the question of why some fields of science converge on a paradigm while others, artlike, remain in a state of constant flux. The answer, he implied, was a matter of choice; scientists within certain fields were simply unwilling to commit themselves to a single paradigm. I suspect Kuhn avoided pursuing this issue because he could not abide the answer. Some fields, such as economics and other social sciences, never adhere for long to a single paradigm because they address questions for which no paradigm will suffice. Fields that achieve consensus, or normalcy, to borrow Kuhn's term, do so because their paradigms correspond to something real in nature, something true.

Finding Feyerabend

To say that the ideas of Popper and Kuhn are flawed is not to say that they cannot serve as useful tools for analyzing science. Kuhn's normal-science model accurately describes what most scientists now do: fill in details, solve relatively trivial puzzles that buttress rather than challenge the prevailing paradigm. Popper's falsification criterion can help to distinguish between empirical science and ironic science. But each philosopher, by pushing his ideas too far, by taking them too seriously, ends up in an absurd, self-contradicting position.

How does a skeptic avoid becoming Karl Popper, pounding the table and shouting that he is *not* dogmatic? Or Thomas Kuhn, trying to communicate precisely what he means when he talks about the impossibility of true communication? There is only one way. One must embrace—even revel in—paradox, contradiction, rhetorical excess. One must acknowledge that skepticism is a necessary but impossible exercise. One must become Paul Feyerabend.

Feyerabend's first and still most influential book, *Against Method*, was published in 1975 and has been translated into 16 languages.[20] It argues that philosophy cannot provide a methodology or rationale for science, since there is no rationale to explain. By analyzing such scientific milestones as Galileo's trial before the Vatican and the development of quantum mechanics, Feyerabend sought to show that there is no logic to science; scientists create and adhere to scientific theories for what are ultimately subjective and even irrational reasons. According to Feyerabend, scientists can and must do whatever is necessary to advance. He summed up his anticredo with the phrase "anything goes." Feyerabend once derided Popper's critical rationalism as "a tiny puff of hot air in the positivistic teacup."[21] He agreed with Kuhn on many points, in particular on the incommensurability of scientific theories, but he argued that science is rarely as normal as Kuhn contended. Feyerabend also accused Kuhn—quite rightly—of avoiding the implications of his own view; he remarked, to Kuhn's dismay, that Kuhn's sociopolitical model of scientific change applied nicely to organized crime.[22]

Feyerabend's penchant for posturing made it all too easy to reduce him to a grab bag of outrageous sound bites. He once likened science to voodoo, witchcraft, and astrology. He defended the right of religious fundamentalists to have their version of creation taught alongside Darwin's theory of evolution in public schools.[23] His entry in the 1991 *Who's Who in America* ended with the following remark: "My life has been the result of accidents, not of goals and principles. My intellectual work forms only an insignificant part of it. Love and personal understanding are much more important. Leading intellectuals with their zeal for objectivity kill these personal elements. They are criminals, not the liberators of mankind."

Feyerabend's Dadaesque rhetoric concealed a deadly serious point: the human compulsion to find absolute truths, however noble, too often culminates in tyranny. Feyerabend attacked science not because he truly believed that it had no more claim to truth than did astrology. Quite the contrary. Feyerabend attacked science because he recognized—and was horrified by—its power, its potential to stamp out the diversity of human thought and culture. He objected to scientific certainty for moral and political, rather than for epistemological, reasons.

At the end of his 1987 book, *Farewell to Reason,* Feyerabend revealed just how deep his relativism ran. He addressed an issue that "has enraged many readers and disappointed many friends—my refusal to condemn even an extreme fascism and my suggestion that it should be allowed to thrive."[24] The point was particularly touchy because Feyerabend had served in the German army during World War II. It would be all too easy, Feyerabend argued, to condemn Nazism, but it was that very moral self-righteousness and certitude that made Nazism possible.

> I say that Auschwitz is an extreme manifestation of an attitude that still thrives in our midst. It shows itself in the treatment of minorities in industrial democracies; in education, education to a humanitarian point of view included, which most of the time consists of turning wonderful young people into colorless and self-righteous copies of their teachers; it becomes manifest in the nuclear threat, the constant increase in the number and power of deadly weapons and the readiness of some so-called patriots to start a war compared with which the holocaust will shrink into insignificance. It shows itself in the killing of nature and of "primitive" cultures with never a thought spent on those thus deprived of meaning for their lives; in the colossal conceit of our intellectuals, their belief that they know precisely what humanity needs and their relentless efforts to recreate people in their own sorry image; in the infantile megalomania of some of our physicians who blackmail their patients with fear, mutilate them and then persecute them with large bills; in the lack of feeling of many so-called searchers for truth who systematically torture animals, study their discomfort and receive prizes for their cruelty. As far as I am concerned there exists no difference between the henchmen of Auschwitz and these "benefactors of mankind."[25]

By the time I tried to track Feyerabend down in 1992, he had retired from the University of California at Berkeley. No one there knew where he was; colleagues assured me that my efforts to find him would be in vain. At Berkeley, he had had a telephone that allowed him to make calls but not receive them. He would accept invitations to conferences and then fail to show up. By mail, he would invite colleagues to visit him. But when they arrived and knocked on the door of his house in the hills overlooking Berkeley, no one would answer.

Later, while skimming *Isis*, a journal of the history and philosophy of science, I found a short review by Feyerabend of a book of essays. The review displayed Feyerabend's talent for one-liners. In response to a denigrating remark the author had made about religion, Feyerabend retorted, "Prayer may not be very efficient when compared to celestial mechanics, but it surely holds its own vis-à-vis some parts of economics."[26]

I called the editor of *Isis* to ask if he knew how I could contact Feyerabend, and he gave me an address near Zurich, Switzerland. I mailed Feyerabend a fawning letter explaining that I wanted to interview him. To my delight, he responded with a chatty, handwritten note saying an interview would be fine. He divided his time between his home in Switzerland and his wife's place in Rome. He enclosed a telephone number for Rome and a photograph of himself wearing an apron and a big grin and standing before a sink full of dishes. The photograph, he explained, "shows me at my favorite activity, washing dishes for my wife in Rome." In mid-October I received another letter from Feyerabend. "This is to tell you that I should be (93%) in New York during the week from October 25 to November 1 and that we might make an interview then. I'll give you a call as soon as I arrive."

So it happened that one chilly night just a few days before Halloween I met Feyerabend at a luxurious Fifth Avenue apartment. The apartment belonged to a former student who had wisely abandoned philosophy for real estate—apparently with some success. She greeted me and led me into her kitchen, where Feyerabend was sitting at a table sipping a glass of red wine. He thrust himself up from a chair and stood crookedly to greet me, as if he suffered from a stiff back; only then did I remember that Feyerabend had been shot in the back and permanently crippled during World War II.

Feyerabend had the energy and angular face of a leprechaun. When we sat down and began talking, he declaimed, sneered, wheedled, and whispered—depending on his point or plot—while whirling his hands like a conductor. Self-deprecation spiced his hubris. He called himself "lazy" and "a bigmouth." When I asked about his position on a certain point, he winced. "I have no position!" he said. "If you have a position, it is always something screwed down." He twisted an invisible screwdriver into the table. "I have opinions that I defend rather vigorously, and then I find out how silly they are, and I give them up!"

Watching this performance with an indulgent smile was Feyerabend's wife, Grazia Borrini, an Italian physicist whose manner was as calm as Feyerabend's was manic. Borrini had taken Feyerabend's class while pursuing a second degree in public health at Berkeley in 1983; they married six years later. Borrini entered the conversation sporadically, for example, after I asked why Feyerabend thought scientists were so infuriated by his writings.

"I have no idea," he said, the very picture of innocence. "Are they?"

Borrini interjected that *she* had been infuriated when she first heard about Feyerabend's ideas from another physicist. "Someone was taking away from me the keys of the universe," she explained. It was only when she read his books herself that she realized Feyerabend's views were much more subtle and astute than his critics claimed. "This is what I think you should want to write about," Borrini said to me, "the great misunderstanding."

"Oh, forget it, he's not my press agent," Feyerabend said.

Like Popper, Feyerabend had been born and raised in Vienna. He studied acting and opera as a teenager. At the same time, he became intrigued with science after attending lectures by an astronomer. Far from seeing his two passions as irreconcilable, Feyerabend envisioned himself becoming both an opera singer and an astronomer. "I would spend my afternoons practicing singing, and my evenings on the stage, and then late at night I would observe the stars," he said.

Then came the war. Germany occupied Austria in 1938, and in 1942 the 18-year-old Feyerabend enlisted in an officers' school. Although he hoped his training period would outlast the war, he ended up in charge of 3,000 men on the Russian front. While fighting against (actually fleeing from) the Russians in 1945, he was shot in the lower back. "I couldn't get up," Feyerabend recalled, "and I still remember this vision: 'Ah, I shall be in a wheelchair rolling up and down between rows of books.' I was very happy."

He gradually recovered the ability to walk, although only with the help of a cane. Resuming his studies at the University of Vienna after the war, he switched from physics to history, grew bored, returned to physics, grew bored again, and finally settled on philosophy. His talent for advancing absurd positions through sheer cleverness led to a growing suspicion that rhetoric rather than truth is crucial for carrying an argument. "Truth itself is a rhetorical term," Feyerabend asserted. Jutting out his chin he intoned, " 'I am searching for the truth.' Oh boy, what a great person."

Feyerabend studied under Popper at the London School of Economics in 1952 and 1953. There he met Lakatos, another brilliant student of Popper. It was Lakatos who, years later, urged Feyerabend to write *Against Method*. "He was my best friend," Feyerabend said of Lakatos. Feyerabend taught at the University of Bristol until 1959 and then moved to Berkeley, where he befriended Kuhn.

Like Kuhn, Feyerabend denied that he was antiscience. What he *did* claim, first, was that there is no scientific method. "That is exactly how it works in the sciences," Feyerabend said. "You have certain ideas that work, and then some new situation turns up and you try something else. It's opportunism. You need a toolbox full of different kinds of tools. Not only a hammer and pins and nothing else." This is what he meant by his much-maligned phrase "anything goes" (and not, as is commonly thought, that one scientific theory is as good as any other). Restricting science to a particular methodology—even one as loosely defined as Popper's falsification scheme or Kuhn's normal science mode—would destroy it, Feyerabend said.

Feyerabend also objected to the claim that science is superior to other modes of knowledge. He was particularly enraged at the tendency of Western states to foist the products of science—whether the theory of evolution, nuclear power plants, or gigantic particle accelerators—on people against their will. "There is separation between state and church," he complained, "but none between state and science!"

Science "provides fascinating stories about the universe, about the ingredients, about the development, about how life came about, and all this stuff," Feyerabend said. But the prescientific "mythmakers," he emphasized, such as singers, court jesters, and bards, earned their living, whereas most modern scientists are supported by taxpayers. "The public is the patron and should have a say in the matter."

Feyerabend added, "Of course I go to extremes, but not to the extremes people accuse me of, namely, throw out science. Throw out the idea science is *first*. *That's* all right. It has to be science from case to case." After all, scientists disagree among themselves on many issues. "People should not take it for granted when a scientist says, 'Everybody has to follow this way.' "

If he was not antiscience, I asked, what did he mean by his statement in *Who's Who* that intellectuals are criminals? "I thought so for a long time,"

Feyerabend responded, "but last year I crossed it out, because there are lots of good intellectuals." He turned to his wife. "I mean, you are an intellectual." "No, I am a physicist," she replied firmly. Feyerabend shrugged. "What does it mean, 'intellectual'? It means people who think about things longer than other people, perhaps. But many of them just ran over other people, saying, 'We have figured it out.' "

Feyerabend noted that many nonindustrialized people had done fine without science. The !Kung bushmen in Africa "survive in surroundings where any Western person would come in and die after a few days," he said. "Now you might say people in this society live much longer, but the question is, what is the quality of life, and that has not been decided."

But didn't Feyerabend realize how annoyed most scientists would be by that kind of statement? Even if the bushmen are happy, they are ignorant, and isn't knowledge better than ignorance? "What's so great about knowledge?" Feyerabend replied. "They are good to each other. They don't beat each other down." People have a perfect right to reject science if they so choose, Feyerabend said.

Did that mean fundamentalist Christians also had the right to have creationism taught alongside the theory of evolution in schools? "I think that 'right' business is a tricky business," Feyerabend responded, "because once somebody has a right they can hit somebody else over the head with that right." He paused. Ideally, he said, children should be exposed to as many different modes of thought as possible so they can choose freely among them. He shifted uneasily in his seat. Sensing an opening, I pointed out that he had not really answered my question about creationism. Feyerabend scowled. "This is a dried-out business. It doesn't interest me very much. Fundamentalism is not the old rich Christian tradition." But American fundamentalists are very powerful, I persisted, and they use the kinds of things Feyerabend says to attack the theory of evolution. "But science has been used to say some people have a low intelligence quotient," he retorted. "So everything is used in many different ways. Science can be used to beat down all sorts of other people."

But shouldn't educators point out that scientific theories are different from religious myths? I asked. "Of course. I would say that science is very popular nowadays," he replied. "But then I have also to let the other side get in as much evidence as possible, because the other side is always given a short presentation." Anyway, so-called primitive people often know far more about their environments, such as the properties of local plants, than do so-called experts. "So to say these people are ignorant is just . . . *this* is ignorance!"

I unloaded my self-refuting question: Wasn't there something contradictory about the way he used all the techniques of Western rationalism to attack Western rationalism? Feyerabend refused to take the bait. "Well, they are just tools, and tools can be used in any way you see fit," he said mildly. "They can't blame me that I use them." Feyerabend seemed bored, distracted. Although he would not admit it, I suspected he was tired of being a radical relativist, of defending the colorful belief systems of the world— astrology, creationism, even fascism!—against the bully of rationalism.

Feyerabend's eyes glittered again, however, when he began talking about a book he was working on. Tentatively titled *The Conquest of Abundance*, it addressed the human passion for reductionism. "All human enterprises," Feyerabend explained, seek to reduce the natural diversity, or "abundance," inherent in reality. "First of all the perceptual system cuts down this abundance or you couldn't survive." Religion, science, politics, and philosophy represent our attempts to compress reality still further. Of course, these attempts to conquer abundance simply create new abundances, new complexities. "Lots of people have been killed, in political wars. I mean, certain opinions are not liked." Feyerabend, I realized, was talking about our quest for *The Answer*, the theory to end all theories.

But *The Answer* will—must—forever remain beyond our grasp, according to Feyerabend. He ridiculed the belief of some scientists that they might someday capture reality in a single theory of everything. "Let them have their belief, if it gives them joy. Let them also give talks about that. 'We touch the infinite!' And some people say"—bored voice—" 'Ya ya, he says he touches the infinite.' And some people say"—thrilled voice—" 'Ya ya! He says he touches the infinite!' But to tell the little children in school, 'Now that is what the truth is,' that is going much too far."

Any description of reality is necessarily inadequate, Feyerabend said. "You think that this one-day fly, this little bit of nothing, a human being— according to today's cosmology!—can figure it all out? This to me seems so crazy! It cannot possibly be true! What they figured out is one particular response to their actions, and this response gives this universe, and the reality that is behind this is laughing! 'Ha ha! They think they have found me out!' "

A medieval philosopher named Dionysius the Pseudo-Areopagite, Feyerabend said, had argued that to see God directly is to see nothing at all. "This to me makes a lot of sense. I can't explain why. This big thing, out of which everything comes, you don't have the means. Your language has been created by dealing with things, chairs, and a few instruments. And just on this tiny earth!" Feyerabend paused, lost in a kind of exaltation. "God is emanations, you know? And they come down and become more and more material. And down, down at the last emanation, you can see a little trace of it and guess at it."

Surprised by this outburst, I asked Feyerabend if he was religious. "I'm not sure," he replied. He had been raised as a Roman Catholic, and then he became a "vigorous" atheist. "And now my philosophy has taken a completely different shape. It can't just be that the universe—Boom!—you know, and develops. It just doesn't make any sense." Of course, many scientists and philosophers have argued that it is pointless to speculate about the sense, or meaning, or purpose of the universe. "But people ask it, so why not? So all this will be stuffed into this book, and the question of abundance will come out of it, and it will take me a long time."

As I prepared to leave, Feyerabend asked how my wife's birthday party had gone the previous night. (I had told Feyerabend about my wife's birthday in the course of arranging my meeting with him.) Fine, I replied. "You're not drifting apart?" Feyerabend persisted, scrutinizing me. "It wasn't the last birthday you will ever celebrate with her?"

Borrini glared at him, aghast. "Why should it be?"

"I don't know!" Feyerabend exclaimed, throwing his hands up. "Because it happens!" He turned back to me. "How long have you been married?" Three years, I said. "Ah, just the beginning. The bad things will come. Just wait 10 years." Now you really sound like a philosopher, I said. Feyerabend laughed. He confessed that he had been married and divorced three times before he met Borrini. "Now for the first time I am so happy to be married."

I said that I had heard his marriage to Borrini had made him more easygoing. "Well, this may be two things," Feyerabend replied. "Getting older you don't have the energy not to be easygoing. And she's certainly made a big difference also." He beamed at Borrini, and she beamed back.

Turning to Borrini, I mentioned the photograph that her husband had sent of himself washing dishes, along with the note saying that performing this chore for his wife was the most important thing he did now.

Borrini snorted. "Once in a blue moon," she said.

"What do you mean, once in a blue moon!" Feyerabend bellowed. "Every day I wash dishes!"

"Once in a blue moon," Borrini repeated firmly. I decided to believe the physicist rather than the relativist.

A little more than a year after my meeting with Feyerabend, the *New York Times* reported, to my dismay, that the "anti-science philosopher" had been killed by a brain tumor.[27] I called Borrini in Zurich to offer my condolences—and, yes, to satisfy my craven journalistic curiosity. She was distraught. It had happened so quickly. Paul had complained of headaches, and then a few months later. . . . Composing herself, she told me, proudly, that Feyerabend had kept working until the end. Just before he died, he finished a draft of his autobiography. (The book, with the typically Feyerabendian title *Killing Time*, was published in 1995. In the final pages, which Feyerabend wrote in his final days, he concluded that love is all that matters in life.)[28] What about the book on abundance? I asked. No, Paul did not have time to finish that, Borrini murmured.

Recalling Feyerabend's excoriation of the medical profession, I could not resist asking, did her husband seek medical treatment for his tumor? Of course, she replied. He had had "total confidence" in his doctors' diagnosis and had been willing to accept any treatment they recommended; the tumor had simply been detected too late for anything to be done.

Why Philosophy Is So Hard

Theocharis and Psimopoulos, the authors of the essay in *Nature* titled "Where Science Has Gone Wrong," were right after all: the ideas of Popper, Kuhn, and Feyerabend are "self-refuting." All skeptics, finally, fall on their own swords. They become what the critic Harold Bloom derided in *The Anxiety of Influence* as "mere rebels." Their most potent argument against scientific truth is historical: given the rapid turnover of scientific theories over the past century or so, how can we be sure that any current theory will endure? Actually, modern science has been much less revolutionary—and more conservative—than the skeptics, and Kuhn in particular, have suggested. Particle physics rests on the firm foundation of quantum mechanics, and modern genetics bolsters rather than undermines Darwin's theory of evolution. The skeptics' historical arguments are much more devastating when turned against philosophy. If science cannot achieve absolute truth, then what standing should be accorded philosophy, which has exhibited much less ability to resolve its problems? Philosophers themselves have recognized their plight. In *After Philosophy: End or Transformation?* published in 1987, fourteen prominent philosophers considered whether their discipline had a future. The consensus was philosophical: maybe, maybe not.[29]

One philosopher who has pondered the "chronic lack of progress" of his calling is Colin McGinn, a native of England who has taught at Rutgers University since 1992. McGinn, when I met him in his apartment on Manhattan's upper West Side in August 1994, was disconcertingly youthful. (Of course, I expect all philosophers to have furrowed brows and hairy ears.) He wore jeans, a white T-shirt, and moccasins. He is a compact man, with a defiantly jutting chin and pale blue eyes; he could pass for Anthony Hopkins's younger brother.

When I solicited McGinn's opinion of Popper, Kuhn, and Feyerabend, his mouth curled in distaste. They were "sloppy," "irresponsible"; Kuhn especially was full of "absurd subjectivism and relativism." Few modern philosophers took his views seriously any more. "I don't think that science is provisional in an interesting sense at all," McGinn asserted. "*Some* of it's provisional, but some of it isn't!" Is the periodic table provisional? Or Darwin's theory of natural selection?

Philosophy, on the other hand, does not achieve this kind of resolution, McGinn said. It does not advance in the sense that "you have this problem and you work on it and you solve it, and then you go on to the next problem." Certain philosophical problems have been "clarified"; certain approaches have fallen out of fashion. But the great philosophical questions—What is truth? Does free will exist? How can we know anything?—are as unresolved today as they ever were. That fact should not be surprising, McGinn remarked, since modern philosophy can be defined as the effort to solve problems lying beyond the scope of empirical, scientific inquiry.

McGinn pointed out that many philosophers in this century—notably Ludwig Wittgenstein and the logical positivists—have simply declared that philosophical problems are pseudoproblems, illusions stemming from language or "diseases of thought." Some of these "eliminativists," in order to solve the mind-body problem, have even denied that consciousness exists. That view "can have political consequences that you might not want to accept," McGinn said. "It ends up reducing human beings to nothing. It pushes you toward extreme materialism, toward behaviorism."

McGinn offered a different, and, he suggested, more palatable explanation: the great problems of philosophy are real, but they are beyond our cognitive ability. We can pose them, but we cannot solve them—any more than a rat can solve a differential equation. McGinn said this idea came to him in a late-night epiphany when he still lived in England; only later did he realize that he had encountered a similar idea in the writings of the linguist Noam Chomsky (whose views will be aired in Chapter 6). In his 1993 book, *Problems in Philosophy*, McGinn suggested that perhaps in a million years philosophers would acknowledge that his prediction was correct.[30] Of course, he told me, philosophers probably would cease struggling to achieve the impossible much sooner.

McGinn suspected that science, too, was approaching a cul-de-sac. "People have great confidence in science and the scientific method," he said, "and it's worked well within its own limits for a few hundred years. But from a larger perspective who's to say that it's going to carry on and conquer everything?" Scientists, like philosophers, are constrained by their cognitive limits. "It's hubris to think we've somehow now got the perfect cognitive instrument in our heads," he said. Moreover, the end of the cold war has removed a major motivation for investment in science, and as the sense of completion in science grows, fewer bright young people will be attracted to scientific careers.

"So it wouldn't surprise me if sometime during the next century people started veering away from studying science as much, except just to learn what they need to know about things, and started to go back into the humanities." In the future, we will look back on science as "a phase, a brilliant phase. People do forget that just 1,000 years ago there was just religious doctrine; that was it." After science has ended, "religion may start to appeal to people again." McGinn, who is a professed atheist, looked rather pleased with himself, and well he might. During our little chat in his airy apartment, with car honks and bus growls and the odor of greasy Chinese food drifting through the window, he had pronounced the impending doom of not just one but two major modes of human knowledge: philosophy and science.

Fearing the Zahir

Of course, philosophy will never really end. It will simply continue in a more overtly ironic, literary mode, like that already practiced by Nietzsche, or Wittgenstein, or Feyerabend. One of my favorite literary philosophers is the Argentinian fabulist Jorge Luis Borges. More than any philosopher I know of, Borges has explored the complex psychological relationship that we have toward the truth. In "The Zahir," Borges told the story of a man who becomes obsessed with a coin he received as change from a store-keeper.[31] The seemingly nondescript coin is a Zahir, an object that is an emblem of all things, of the mystery of existence. A Zahir can be a compass, a tiger, a stone, anything. Once beheld, it cannot be forgotten. It grips the mind of the beholder until all other aspects of reality become insignificant, trivial.

At first, the narrator struggles to free his mind of the Zahir, but he eventually accepts his fate. "I shall pass from thousands of apparitions to one alone: from a very complex dream to a very simple dream. Others will dream that I am mad, and I shall dream of the Zahir. And when everyone dreams of the Zahir day and night, which will be a dream and which a reality, the earth or the Zahir?"[32] The Zahir, of course, is *The Answer*, the secret of life, the theory to end all theories. Popper, Kuhn, and Feyerabend tried to protect us from *The Answer* with doubt and reason, Borges with terror.

CHAPTER THREE

The End of Physics

There are no more dedicated, not to say obsessive, seekers of *The Answer* than modern particle physicists. They want to show that all the complicated things of the world are really just manifestations of one thing. An essence. A force. A loop of energy wriggling in a 10-dimensional hyperspace. A sociobiologist might suspect that a genetic influence lurks behind this reductionist impulse, since it seems to have motivated thinkers since the dawn of civilization. God, after all, was conceived by the same impulse.

Einstein was the first great modern *Answer* seeker. He spent his later years trying to find a theory that would unify quantum mechanics with his theory of gravity, general relativity. To him, the purpose of finding such a theory was to determine whether the universe was inevitable or, as he put it, "whether God had any choice in creating the world." But Einstein, no doubt believing that science made life meaningful, also suggested that no theory could be truly final. He once said of his own theory of relativity, "[It] will have to yield to another one, for reasons which at present we do not yet surmise. I believe that the process of deepening the theory has no limits."[1]

Most of Einstein's contemporaries saw his efforts to unify physics as a product of his dotage and quasi-religious tendencies. But in the 1970s, the dream of unification was revived by several advances. First, physicists showed that just as electricity and magnetism are aspects of a single force, so electromagnetism and the weak nuclear force (which governs certain kinds of nuclear decay) are manifestations of an underlying "electroweak" force. Researchers also developed a theory for the strong nuclear force, which grips protons and neutrons together in the nuclei of atoms. The theory, called quantum chromodynamics, posits that protons and neutrons are composed of even more elementary particles, called quarks. Together, the electroweak theory and quantum chromodynamics constitute the standard model of particle physics.

Emboldened by this success, workers forged far beyond the standard model in search of a deeper theory. Their guide was a mathematical property called symmetry, which allows the elements of a system to undergo transformations—analogous to rotation or reflection in a mirror—without being fundamentally altered. Symmetry became the *sine qua non* of particle physics. In search of theories with deeper symmetries, theorists began to jump to higher dimensions. Just as an astronaut rising above the two-dimensional plane of the earth can more directly apprehend its global symmetry, so can theorists discern the more subtle symmetries underlying particle interactions by viewing them from a higher-dimensional standpoint.

One of the most persistent problems in particle physics stems from the definition of particles as points. In the same way that division by zero yields an infinite and hence meaningless result, so do calculations involving pointlike particles often culminate in nonsense. In constructing the standard model, physicists were able to sweep these problems under the rug. But Einsteinian gravity, with its distortions of space and time, seemed to demand an even more radical approach.

In the early 1980s, many physicists came to believe that superstring theory represented that approach. The theory replaced pointlike particles with minute loops of energy that eliminated the absurdities arising in calculations. Just as vibrations of violin strings give rise to different notes, so could the vibrations of these strings generate all the forces and particles of the physical realm. Superstrings could also banish one of the bugbears of particle physics: the possibility that there is no ultimate foundation for physical reality but only an endless succession of smaller and smaller particles, nestled inside each other like Russian dolls. According to superstring theory, there is a fundamental scale beyond which all questions concerning space and time become meaningless.

The theory suffers from several problems, however. First, there seem to be countless possible versions, and theorists have no way of knowing which one is correct. Moreover, superstrings are thought to inhabit not only the four dimensions in which we live (the three dimensions of space plus time), but also six extra dimensions that are somehow "compactified," or rolled up into infinitesimal balls, in our universe. Finally, the strings are as small in comparison to a proton as a proton is in comparison to the solar system. They are more distant from us, in a sense, than are the quasars that lurk at the farthest edge of the visible universe. The superconducting supercollider, which was to take physicists much deeper into the microrealm than had any previous accelerator, would have been 54 miles in circumference. To probe the realm superstrings are thought to inhabit, physicists would have to build a particle accelerator 1,000 light-years around. (The entire solar system is only one light-*day* around.) And not even an accelerator that size could allow us to see the extra dimensions where superstrings dance.

Glashow's Gloom

One of the pleasures of being a science writer is feeling superior to run-of-the-mill newshounds. The most primitive species of reporter, to my mind, is the type who tracks down a woman who has watched her only son stabbed to death by a crazed crackhead and asks, "How do you feel?" Yet in the fall of 1993 I found myself stuck with a similar assignment. I was just beginning an article on the future of particle physics when the U.S. Congress killed, once and for all, the superconducting supercollider. (Contractors had already spent more than $2 billion and dug a tunnel in Texas 15 miles long.) Over the next few weeks I had to confront particle physicists who had just seen their brightest hope for the future brutally aborted and ask, "How do you feel?"

The gloomiest place I visited was the department of physics at Harvard University. The head of the department was Sheldon Glashow, who had shared a Nobel Prize with Steven Weinberg and Abdus Salam for developing the electroweak portion of the standard model. In 1989, Glashow had spoken, along with the biologist Gunther Stent, at the symposium titled "The End of Science?" at Gustavus Adolphus College. Glashow offered a spirited rebuttal of the meeting's "absurd" premise, that philosophical skepticism was eroding belief in science as a "unified, universal, objective endeavor." Does anyone really doubt the existence of the moons of Jupiter, which Galileo discovered centuries ago? Does anyone really doubt the modern theory of disease? "Germs are seen and killed," Glashow declared, "not imagined and unimagined."

Science is "certainly slowing down," Glashow conceded, but not because of attacks from ignorant, antiscientific sophists. His own field of particle physics "is threatened from an entirely different direction: from its very success." The last decade of research has generated countless confirmations of the standard model of particle physics "but has revealed not the slightest flaw, not the tiniest discrepancy. . . . We have no experimental hint or clue that could guide us to build a more ambitious theory." Glashow added the obligatory hopeful coda: "Nature's road has often seemed impassable, but we have always overcome."[2]

In other statements, Glashow has not hewed to this blithe optimism. He was once a leader in the quest for a unified theory. In the 1970s, he proposed several such theories, although none as ambitious as superstring theory. But with the advent of superstrings, Glashow became disillusioned with the quest for unification. Those working on superstrings and other unified theories were not doing physics at all any more, Glashow contended, because their speculations were so far beyond any possible empirical test. Glashow and a colleague complained in one essay that "contemplation of superstrings may evolve into an activity as remote from conventional particle physics as particle physics is from chemistry, to be conducted at schools of divinity by future equivalents of medieval theologians." They added that "for the first time since the Dark Ages, we can see how our noble search may end, with faith replacing science once again."[3] When particle physics passes beyond the realm of the empirical, Glashow seemed to be suggesting, it may succumb to skepticism, to relativism, after all.

I interviewed Glashow at Harvard in November 1993, shortly after the demise of the superconducting supercollider. His dimly lit office, lined with dark, heavily varnished bookcases and cabinets, was as solemn as a funeral parlor. Glashow himself, a large man chewing restlessly on a cold cigar stub, seemed slightly incongruous there. He had the snowy, tousled hair that seems de rigueur for Nobel laureates in physics, and his glasses were as thick as telescope lenses. Yet one could still detect, beneath the patina of the Harvard professor, the tough, fast-talking New York kid that Glashow had once been.

Glashow was devastated by the death of the supercollider. Physics, he emphasized, cannot proceed on pure thought alone, in spite of what superstring enthusiasts said. Superstring theory "hasn't gotten anywhere despite all the hoopla," he grumbled. More than a century ago some physicists tried to invent unified theories; they failed, of course, because they knew nothing about electrons or protons or neutrons or quantum mechanics. "Now, are we so arrogant as to believe we have all the experimental information we need right now to construct that holy grail of theoretical physics, a unified theory? I think not. I think certainly there are surprises that natural phenomena have in store for us, and we're not going to find them unless we look."

But isn't there much to do in physics besides unification? "Of course there is," Glashow replied sharply. Astrophysics, condensed-matter physics, and even subfields within particle physics are not concerned with unification. "Physics is a very large house filled with interesting puzzles," he said (using Thomas Kuhn's term for problems whose solutions merely reinforce the prevailing paradigm). "Of *course* there will be things done. The question is whether we're getting anywhere toward this holy grail." Glashow believed that physicists would continue to seek "some little interesting tidbit someplace. Something amusing, something new. But it's not the same as the quest as I was fortunate enough to know it in my professional lifetime."

Glashow could not muster much optimism concerning the prospects for his field, given the politics of science funding. He had to admit that particle physics was not terribly useful. "Nobody can make the claim that this kind of research is going to produce a practical device. That would just be a lie. And given the attitude of governments today, the type of research that I fancy doesn't have a very good future."

In that case, could the standard model be the final theory of particle physics? Glashow shook his head. "Too many questions left unanswered," he said. Of course, he added, the standard model would be final in a practical sense if physicists could not forge beyond it with more powerful accelerators. "There will be the standard theory, and that will be the last chapter in the elementary physics story." It is always possible that someone will find a way of generating extremely high energies relatively cheaply. "Maybe someday it will get done. Someday, someday, someday."

The question is, Glashow continued, what will particle physicists do while they are waiting for that someday to come? "I guess the answer is going to be that the [particle-physics] establishment is going to do boring things, futzing around until something becomes available. But they would never admit that it's boring. Nobody will say, 'I do boring things.' " Of course, as the field becomes less interesting and funding dwindles, it will cease attracting new talent. Glashow noted that several promising graduate students had just left Harvard for Wall Street. "Goldman Sachs in particular discovered that theoretical physicists are very good people to have."

The Smartest Physicist of Them All

One reason that superstring theory became so popular in the mid-1980s was that a physicist named Edward Witten decided it represented the best hope for the future of physics. I first saw Witten in the late 1980s when I was eating lunch with another scientist in the cafeteria of the Institute for Advanced Study in Princeton. A man walked by our table holding a tray of food. He had a lantern jaw and a strikingly high forehead, bounded by thick black glasses across the bottom and thick black hair across the top. "Who's that?" I asked my companion. "Oh, that's Ed Witten," my lunchmate replied. "He's a particle physicist."

A year or two later, making idle chitchat between sessions at a physics conference, I asked a number of attendees: Who is the smartest physicist of them all? Several names kept coming up, including the Nobel laureates Steven Weinberg and Murray Gell-Mann. But the name mentioned most often was Witten's. He seemed to evoke a special kind of awe, as though he belonged to a category unto himself. He is often likened to Einstein; one colleague reached even further back for a comparison, suggesting that Witten possessed the greatest mathematical mind since Newton.

Witten may also be the most spectacular practitioner of naive ironic science I have ever encountered. Naive ironic scientists possess an exceptionally strong faith in their scientific speculations, in spite of the fact that those speculations cannot be empirically verified. They believe they do not invent their theories so much as they discover them; these theories exist independently of any cultural or historical context and of any particular efforts to find them.

A naive ironic scientist—like a Texan who thinks everyone but Texans has an accent—does not acknowledge that he or she has adopted any philosophical stance at all (let alone one that might be described as ironic).

Such a scientist is just a conduit through which truths pass from the Platonic realm to the world of flesh; background and personality are irrelevant to the scientific work. Thus Witten, when I called him to request an interview, tried to dissuade me from writing about him. He told me that he abhorred journalism focusing on scientists' personalities, and at any rate, many other physicists and mathematicians were much more interesting than he. Witten had been upset by a profile published in the *New York Times Magazine* in 1987, which implied that he had invented superstring theory.[4] Actually, Witten informed me, he had played virtually no role in the creation of superstring theory; he had simply helped to develop it and promote it after it had been discovered.

Every science writer occasionally runs into subjects who genuinely do not want attention from the media, who simply want to be left alone to do their work. What these scientists often fail to realize is that this trait makes them more enticing. Intrigued by Witten's apparently sincere shyness, I persisted in seeking an interview. Witten asked me to send him some things I had written. Stupidly, I included a profile of Thomas Kuhn published in *Scientific American*. Finally, Witten agreed to let me come down to talk to him, but he would give me two hours and not a minute more; I would have to leave at 12 noon, sharp. When I arrived, he immediately began lecturing me on my shoddy journalistic ethics. I had done society a disservice in repeating Thomas Kuhn's view that science is an arational (not irrational) process that does not converge on the truth. "You should be concentrating on serious and substantive contributions to the understanding of science," Witten said. Kuhn's philosophy "isn't taken very seriously except as a debating standard, even by its proponents." Did Kuhn go to a doctor when he was sick? Did he have radial tires on his car? I shrugged and guessed that he probably did. Witten nodded triumphantly. That proved, he declared, that Kuhn believed in science and not in his own relativistic philosophy.

I said that whether or not one agreed with Kuhn's views, they were influential, and provocative, and that one of my aims as a writer was not only to inform readers but also to provoke them. "Aim to report on some of the *truths* that are being discovered, rather than aiming to provoke. *That* should be the aim of a science writer," Witten said sternly. I tried to do both, I replied. "Well, that's a pretty feeble response," Witten said. "Provoking people, or stimulating them intellectually, should be a *by-product* of reporting on some of the truths that are being discovered." This is another mark of the naive ironic scientist: when he or she says "truth," there is never any ironic inflection, no smile; the word is implicitly capitalized. As penance for my journalistic sins, Witten suggested, I should write profiles of five mathematicians in a row; if I did not know which mathematicians were worthy of such attention, Witten would recommend some. (Witten did not realize that he was providing fodder for those who claimed he was less a physicist than a mathematician.)

Since high noon was approaching, I tried to turn the interview toward Witten's career. He refused to answer any "personal" questions, such as what his major had been in college or whether he had considered other careers before becoming a physicist; his history was unimportant. I knew from background reading that although Witten was the son of a physicist and had always enjoyed the subject, he had graduated from Brandeis College in 1971 with a degree in history and plans to become a political journalist. He succeeded in publishing articles in the *New Republic* and the *Nation*. He nonetheless soon decided that he lacked the "common sense" for journalism (or so he told one reporter); he entered Princeton to study physics and obtained his doctorate in 1976.

Witten picked up the story from there. In telling me about his work in physics, he slipped into a highly abstract, impersonal mode of speech. He was reciting, not talking, giving me the history of superstrings, emphasizing not his own role but that of others. He spoke so softly that I worried that my tape recorder wouldn't pick him up over the air conditioner. He paused frequently—for 51 seconds at one point—casting his eyes down and squeezing his lips together like a bashful teenager. He seemed to be striving for the same precision and abstraction in his speech that he achieved in his treatises on superstrings. Now and then—for no reason that I could discern—he broke into convulsive, hiccuping laughter as some private joke flitted through his consciousness.

Witten made a name for himself in the mid-1970s with incisive but fairly conventional papers on quantum chromodynamics and the electroweak force. He learned of superstring theory in 1975, but his initial efforts to understand it were stymied by the "opaque" literature. (Yes, even the smartest person in the world had a hard time understanding superstring theory.) In 1982, however, a review paper by John Schwarz, one of the theory's pioneers, helped Witten to grasp a crucial fact: Superstring theory does not simply allow for the possibility of gravity; it demands that gravity exist. Witten called this realization "the greatest intellectual thrill of my life." Within a few years, any doubts that Witten had had about the theory's potential vanished. "It was clear that if I didn't spend my life concentrating on string theory, I would simply be missing my life's calling," he said. He began publicly proclaiming the theory a "miracle" and predicting that it would "dominate physics for the next 50 years." He also generated a flood of papers on the theory. The 96 articles that Witten produced from 1981 through 1990 were cited by other physicists 12,105 times; no other physicist in the world approached this level of influence.[5]

In his early papers, Witten concentrated on creating a superstring model that was a reasonable facsimile of the real world. But he became increasingly convinced that the best way to achieve that goal was to uncover the theory's "core geometric principles." These principles, he said, might be analogous to the non-Euclidean geometry that Einstein employed to construct his theory of general relativity. Witten's pursuit of these ideas led him deep into topology, which is the study of the fundamental geometric properties of objects, regardless of their particular shape or size. In the eyes of a topologist, a doughnut and a single-handled coffee mug are equivalent, because each has only one hole; one object can be transformed into the other without any tearing. A doughnut and a banana are not equivalent, because one would have to tear the doughnut to mold it into a banana-shaped object. Topologists are particularly fond of determining whether seemingly dissimilar knots can actually be transformed into each other without being cut. In the late 1980s Witten created a technique—which borrowed from both topology and quantum field theory—that allows mathematicians to uncover deep symmetries between hideously tangled, higher-dimensional knots. As a result of his finding, Witten won the 1990 Fields Medal, the most prestigious prize in *mathematics*. Witten called the achievement his "single most satisfying piece of work."

I asked Witten how he responded to the claims of critics that superstring theory is not testable and therefore is not really physics at all. Witten replied that the theory had predicted gravity. "Even though it is, properly speaking, a postprediction, in the sense that the experiment was made before the theory, the fact that gravity is a consequence of string theory, to me, is one of the greatest theoretical insights ever."

He acknowledged, even emphasized, that no one had truly fathomed the theory, and that it might be decades before it yielded a precise description of nature. He would not predict, as others had, that superstring theory might bring about the end of physics. Nevertheless, he was serenely confident that it would eventually yield a profound new understanding of reality. "Good wrong ideas are extremely scarce," he said, "and good wrong ideas that even remotely rival the majesty of string theory have never been seen." When I continued to press Witten on the issue of testability, he grew exasperated. "I don't think I've succeeded in conveying to you its wonder, its incredible consistency, remarkable elegance, and beauty." In other words, superstring theory is too beautiful to be wrong.

Witten then revealed just how strong his faith was. "Generally speaking, all the really great ideas of physics are really spin-offs of string theory," Witten began. "Some of them were discovered first, but I consider that a mere accident of the development on planet earth. On planet earth, they were discovered in this order." Stepping up to his chalkboard, he wrote down general relativity, quantum field theory, superstrings, and supersymmetry (a concept that serves a vital role in superstring theory). "But I don't believe, if there are many civilizations in the universe, that those four ideas were discovered in that order in each civilization." He paused. "I do believe, by the way, that those four ideas were discovered in any advanced civilization."

I could not believe my good fortune. Who's being provocative now? I asked. "I'm not being provocative," Witten retorted. "I'm being provocative in the same way as someone who says the sky is blue is being provocative, if there is a writer somewhere who has said that the sky has pink polka dots."

Particle Aesthetics

In the early 1990s, when superstring theory was still relatively novel, several physicists wrote popular books about its implications. In *Theories of Everything*, the British physicist John Barrow argued that Gödel's incompleteness theorem undermines the very notion of a *complete* theory of nature.[6] Gödel established that any moderately complex system of axioms inevitably raises questions that cannot be answered by the axioms. The implication is that any theory will always have loose ends. Barrow also pointed out that a unified theory of particle physics would not really be a theory of everything, but only a theory of all particles and forces. The theory would have little or nothing to say about phenomena that make our lives meaningful, such as love or beauty.

But Barrow and other analysts at least granted that physicists might achieve a unified theory. That assumption was challenged in *The End of Physics*, written by physicist-turned-journalist David Lindley.[7] Physicists working on superstring theory, Lindley contended, were no longer doing physics because their theories could never be validated by experiments, but only by subjective criteria, such as elegance and beauty. Particle physics, Lindley concluded, was in danger of becoming a branch of aesthetics.

The history of physics supports Lindley's prognosis. Previous theories of physics, however seemingly bizarre, won acceptance among physicists and even the public not because they made sense; rather, they offered predictions that were borne out—often in dramatic fashion—by observations. After all, even Newton's version of gravity violates common sense. How can one thing tug at another across vast spans of space? John Maddox, the editor of *Nature*, once argued that if Newton submitted his theory of gravity to a journal today, it would almost certainly be rejected as too preposterous to believe.[8] Newton's formalism nonetheless provided an astonishingly accurate means of calculating the orbits of planets; it was too effective to deny.

Einstein's theory of general relativity, with its malleable space and time, is even more bizarre. But it became widely accepted as true after observations confirmed his prediction about how gravity would bend light passing around the sun. Likewise, physicists do not believe quantum mechanics because it explains the world, but because it predicts the outcome of experiments with almost miraculous accuracy. Theorists kept predicting new particles and other phenomena, and experiments kept bearing out those predictions.

Superstring theory is on shaky ground indeed if it must rely on aesthetic judgments. The most influential aesthetic principle in science was set forth by the fourteenth-century British philosopher William of Occam. He argued that the best explanation of a given phenomenon is generally the simplest, the one with the fewest assumptions. This principle, called Occam's razor, was the downfall of the Ptolemaic model of the solar system in the Middle Ages. To show that the earth was the center of the solar system, the astronomer Ptolemy was forced to argue that the planets traced elaborately spiraling epicycles around the earth. By assuming that the sun and not the earth was the center of the solar system, later astronomers eventually could dispense with epicycles and replace them with much simpler elliptical orbits.

Ptolemy's epicycles seem utterly reasonable when compared to the undetected—and undetectable—extra dimensions required by superstring theory. No matter how much superstring theorists assure us of the theory's mathematical elegance, the metaphysical baggage it carries with it will prevent it from winning the kind of acceptance—among either physicists or laypeople—that general relativity or the standard model of particle physics have.

Let's give superstring believers the benefit of the doubt, if only for a moment. Let's assume that some future Witten, or even Witten himself, finds an infinitely pliable geometry that accurately describes the behavior of all known forces and particles. In what sense will such a theory explain the world? I have talked to many physicists about superstrings, and none has been able to help me understand what, exactly, a superstring *is*. As far as I can tell, it is neither matter nor energy; it is some kind of mathematical ur-stuff that generates matter and energy and space and time but does not itself correspond to anything in our world.

Good science writers will no doubt make readers think they understand such a theory. Dennis Overbye, in *Lonely Hearts of the Cosmos*, one of the best books ever written on cosmology, imagines God as a cosmic rocker, bringing the universe into being by flailing on his 10-dimensional superstring guitar.[9] (One wonders, is God improvising, or following a score?) The true meaning of superstring theory, of course, is embedded in the theory's austere mathematics. I once heard a professor of literature liken James Joyce's gobbledygookian tome *Finnegans Wake* to the gargoyles atop the cathedral of Notre Dame, built solely for God's amusement. I suspect that if Witten ever finds the theory he so desires, only he—and God, perhaps—will truly appreciate its beauty.

Nightmares of a Final Theory

With his crab-apple cheeks, vaguely Asian eyes, and silver hair still tinged with red, Steven Weinberg resembles a large, dignified elf. He would make an excellent Oberon, king of the fairies in *A Midsummer Night's Dream*. And like a fairy king, Weinberg has demonstrated a powerful affinity for the mysteries of nature, an ability to discern subtle patterns within the froth of data streaming from particle accelerators. In his 1993 book, *Dreams of a Final Theory*, he managed to make reductionism sound romantic. Particle physics is the culmination of an epic quest, "the ancient search for those principles that cannot be explained in terms of deeper principles."[10] The force compelling science, he pointed out, is the simple question, why? That question has led physicists deeper and deeper into the heart of nature. Eventually, he contended, the convergence of explanations down to simpler and simpler principles would culminate in a final theory. Weinberg suspected that superstrings might lead to that ultimate explanation.

Weinberg, like Witten and almost all particle physicists, has a profound faith in the power of physics to achieve absolute truth. But what makes Weinberg such an interesting spokesperson for his tribe is that he, unlike Witten, is acutely aware that his faith is just that, a faith; Weinberg knows that he is speaking with a philosophical accent. If Edward Witten is a philosophically naive scientist, Weinberg is an extremely sophisticated one—too sophisticated, perhaps, for the good of his own field.

I first met Weinberg in New York in March 1993—during the halcyon era before the supercollider was killed—at a dinner held to celebrate the publication of *Dreams of a Final Theory*. He was in an expansive mood, dispensing jokes and anecdotes about famous colleagues and wondering what it would be like to chat with the talk-show host Charlie Rose later that night. Eager to impress the great Nobel laureate, I began name-dropping. I mentioned that Freeman Dyson had recently told me that the whole notion of a final theory was a pipe dream.

Weinberg smiled. The majority of his colleagues, he assured me, believed in a final theory, although many of them preferred to keep that belief private. I dropped another name, Jack Gibbons, whom the newly elected Bill Clinton had just named science advisor. I had recently interviewed Gibbons, I said, and Gibbons had hinted that the United States alone might not be able to afford the supercollider. Weinberg scowled and shook his head, muttering something about society's disturbing lack of appreciation for the intellectual benefits of basic research.

The irony was that Weinberg himself, in *Dreams of a Final Theory*, offered little or no argument as to why society *should* support further research in particle physics. He was careful to acknowledge that neither the superconducting supercollider nor any other earthly accelerator could provide direct confirmation of a final theory; physicists would eventually have to rely on mathematical elegance and consistency as guides. Moreover, a final theory might have no practical value. Weinberg's most extraordinary admission was that a final theory might not reveal the universe to be meaningful in human terms. Quite the contrary. He reiterated an infamous comment made in an earlier book: "The more the universe seems comprehensible, the more it seems pointless."[11] Although the comment had "dogged him ever since," Weinberg refused to back away from it. Instead he elaborated on the remark: "As we have discovered more and more fundamental physical principles they seem to have less and less to do with us."[12] Weinberg seemed to be acknowledging that all our whys would eventually culminate in a because. His vision of the final theory evoked *The Hitchhiker's Guide to the Galaxy* by Douglas Adams. In this science fiction comedy, published in 1980, scientists finally discover the answer to the riddle of the universe, and the answer is . . . 42. (Adams was practicing philosophy of science in an overtly literary mode.)

The superconducting supercollider was dead and buried when I met Weinberg again in March 1995 at the University of Texas at Austin. His spacious office was cluttered with periodicals that testified to the breadth of his interests, including *Foreign Affairs, Isis, Skeptical Inquirer,* and *American Historical Review,* as well as physics journals. Along one wall ran a chalkboard laced with the obligatory mathematical scribbles. Weinberg spoke with what seemed to be considerable effort. He kept sighing, grimacing, squeezing his eyes shut and rubbing them, even as his deep, sonorous voice rolled forward. He had just eaten lunch and was probably experiencing postprandial fatigue. But I preferred to think he was brooding over the tragic dilemma of particle physicists: they are damned if they achieve a final theory and damned if they don't.

It is a "terrible time for particle physics," Weinberg admitted. "There's never been a time when there's been so little excitement in the sense of experiments suggesting really new ideas or theories being able to make new and qualitatively different kinds of predictions that are then borne out by experiments." With the supercollider dead and plans for other new accelerators in the United States stalled for lack of funding, the prospects for the field were gloomy. Oddly enough, brilliant students were still entering the field, students "better than we deserve, probably," Weinberg added.

Although he shared Witten's belief that physics moves toward absolute truth, Weinberg was acutely aware of the philosophical difficulties of defending this position. He recognized that "the techniques by which we decide on the acceptance of physical theories are extremely subjective." It would always be possible for clever philosophers to make a case that the particle physicists "are just making it up as they go along." (In *Dreams of a Final Theory*, Weinberg even confessed to having a fondness for the writings of the philosophical anarchist Paul Feyerabend.) On the other hand, Weinberg told me, the standard model of particle physics, "whatever the aesthetics were, [has] by now been tested as few theories have been, and it really works. If in fact it was just a social construct it would have fallen apart long before this."

Weinberg realized that physicists would never be able to *prove* a theory final in the same way that mathematicians prove theorems; but if the theory accounted for all experimental data—the masses of all particles, the strengths of all forces—physicists would eventually cease questioning it. "I don't feel I was put here to be sure of anything," Weinberg said. "A lot of philosophy of science going back to the Greeks has been poisoned by the search for certainty, which seems to me a false search. Science is too much fun to sit around wringing our hands because we're not certain about things."

Even as we spoke, Weinberg suggested, someone might be posting the final, correct version of superstring theory on the Internet. "If *she*,"— Weinberg added, with the slightest pause and emphasis on the *she*—"got results that agreed with experiment, then you would say, 'That's it,' " even if researchers could never provide direct evidence of the strings themselves or the extra dimensions they supposedly inhabit; after all, the atomic theory of matter was accepted because it worked, not because experimenters could make pictures of atoms. "I agree strings are much farther away from direct perception than atoms, and atoms are much farther away from direct perception than chairs, but I don't see any philosophical discontinuity there."

There was little conviction in Weinberg's voice. Deep down, he surely knew that superstring theory *did* represent a discontinuity in physics, a leap beyond any conceivable empirical test. Abruptly, he rose and began prowling around the room. He picked up odd objects, fondled them distractedly, put them down, while continuing to speak. He reiterated his belief that a final theory of physics would represent the most fundamental possible achievement of science—the bedrock of all other knowledge. To be sure, some complex phenomena, such as turbulence, or economics, or life, require their own special laws and generalizations. But if you ask why those principles are true, Weinberg added, that question takes you down toward the final theory of physics, on which everything rests. "That's what makes science a hierarchy. And it *is* a hierarchy. It's not just a random net."

Many scientists cannot abide hearing that truth, Weinberg said, but there is no escaping it. "Their final theory is what our final theory explains." If neuroscientists ever explain consciousness, for example, they will explain it in terms of the brain, "and the brain is what it is because of historical accidents and because of universal principles of chemistry and physics." Science will certainly continue after a final theory, perhaps forever, but it will have lost something. "There will be a sense of sadness" about the achievement of a final theory, Weinberg said, since it will bring to a close the great quest for fundamental knowledge.

As Weinberg continued speaking, he seemed to portray the final theory in increasingly negative terms. Asked whether there would ever be such a thing as applied superstring theory, Weinberg grimaced. (In the 1994 book *Hyperspace*, the physicist Michio Kaku foresaw a day when advances in superstring theory would allow us to visit other universes and travel through time.)[13] Weinberg cautioned that "the sands of scientific history are white with the bones of people" who failed to foresee applications of developments in science, but applied superstring theory was "hard to conceive."

Weinberg also doubted that the final theory would resolve all the notorious paradoxes posed by quantum mechanics. "I tend to think these are just puzzles in the way we talk about quantum mechanics," Weinberg said. One way to eliminate these puzzles, he added, would be to adopt the many-worlds interpretation of quantum mechanics. Proposed in the 1950s, this interpretation attempts to explain why the act of observation by a physicist seems to force a particle such as an electron to choose only one path out of the many allowed by quantum mechanics. According to the many-worlds interpretation, the electron actually follows all possible paths, but in separate universes. This explanation does have its troubling aspects, Weinberg conceded. "There may be another parallel time track where John Wilkes Booth missed Lincoln and . . ." Weinberg paused. "I sort of hope that whole problem will go away, but it may not. That may be just the way the world is."

Is it too much to ask for a final theory to make the world intelligible? Before I could finish the question, Weinberg was nodding. "Yes, it's too much to ask," he replied. The proper language of science is mathematics, he reminded me. A final theory "has to make the universe appear plausible and somehow or other recognizably logical to people who are trained in that language of mathematics, but it may be a long time before that makes sense to other people." Nor will a final theory provide humanity with any guidance in conducting its affairs. "We've learned to absolutely disentangle value judgments from truth judgments," Weinberg said. "I don't see us going back to reconnect them." Science "can certainly help you find out what the consequences of your actions are, but it can't tell you what consequences you ought to wish for. And that seems to me to be an absolute distinction."

Weinberg had little patience for those who suggest that a final theory will reveal the purpose of the universe, or "the mind of God," as Stephen Hawking once put it. Quite the contrary. Weinberg hoped that a final theory would eliminate the wishful thinking, mysticism, and superstition that pervades much of human thought, even among physicists. "As long as we don't know the fundamental rules," he said, "we can hope that we'll find something like a concern for human beings, say, or some guiding divine plan built into the fundamental rules. But when we find out that the fundamental rules of quantum mechanics and some symmetry principles are very impersonal and cold, then it'll have a very demystifying effect. At least that's what I'd like to see."

His face hardening, Weinberg continued: "I certainly would not disagree with people who say that physics in my style or the Newtonian style has produced a certain disenchantment. But if that's the way the world is, it's better we find out. I see it as part of the growing up of our species, just like the child finding out there is no tooth fairy. It's better to find out there is no tooth fairy, even though a world with tooth fairies in it is somehow more delightful."

Weinberg was well aware that many people hungered for a different message from physics. In fact, earlier that day he had heard that Paul Davies, an Australian physicist, had received a million-dollar prize for "advancing public understanding of God or spirituality." Davies had written numerous books, notably *The Mind of God*, suggesting that the laws of physics reveal a plan underlying nature, a plan in which human consciousness may play a central role.[14] After telling me about Davies's prize, Weinberg chuckled mirthlessly. "I was thinking of cabling Davies and saying, 'Do you know of any organization that is willing to offer a million-dollar prize for work showing that there is no divine plan?' "

In *Dreams of a Final Theory*, Weinberg dealt rather harshly with all this talk of divine plans. He raised the embarrassing issue of human suffering. What kind of plan is it that allows the Holocaust, and countless other evils, to happen? What kind of planner? Many physicists, intoxicated by the power of their mathematical theories, have suggested that "God is a geometer." Weinberg retorted, in effect, that he does not see why we should be interested in a God who seems so little interested in us, however good he is at geometry.

I asked Weinberg what gave him the fortitude to sustain such a bleak (and in my view, accurate) vision of the human condition. "I sort of enjoy my tragic view," he replied with a little smile. "After all, which would you rather see, a tragedy or—" he hesitated, his smile fading. "Well, some people would prefer to see a comedy. But . . . I think the tragic view adds a certain dimension to life. Anyway, it's the best we have." He stared out his office window, brooding. Fortunately, perhaps, for Weinberg, his view did not include the infamous tower from which a deranged University of Texas student, Charles Whitman, shot 14 people to death in 1966. Weinberg's office overlooks a graceful Gothic church that serves as the university's theological seminary. But Weinberg did not seem to be looking at the church—or, for that matter, at anything else in the material world.

No More Surprises

Even if society musters the will and the money to build larger accelerators and thereby keep particle physics alive—at least temporarily—how likely is it that physicists will learn something as truly new and surprising as, say, quantum mechanics? Not very, according to Hans Bethe. A professor at Cornell University, Bethe won a Nobel Prize in 1967 for his work on the carbon cycle in stellar fusion; in other words, he showed how stars shine. He also headed the theoretical division of the Manhattan Project during World War II. In that capacity he made what was arguably the most important calculations in the history of the planet. Edward Teller (who, ironically, later became the scientific community's most avid booster of nuclear weapons) had done some calculations suggesting that the fireball from an atomic blast might ignite the earth's atmosphere, triggering a conflagration that would consume the entire world. The scientists studying Teller's conjecture took it very seriously; after all, they were exploring terra incognita. Bethe then examined the problem and made his own calculations. He determined that Teller was wrong; the fireball would not spread.[15]

No one should have to carry out calculations on which the fate of the earth depends. But if someone must, Bethe would be my choice. He exudes wisdom and gravitas. When I inquired whether he had had any lingering doubts, in the moments before the bomb was detonated at Alamogordo, about what would happen, he shook his head. No, he replied. His only concern had been whether the ignition device would work properly. There was not a trace of braggadocio in Bethe's response. He had done the calculations, and he trusted them. (One wonders whether even Edward Witten would entrust the fate of the earth to a prediction based on superstrings.)

When I asked him about the future of his field, Bethe said there were still many open questions in physics, including ones raised by the standard model. Moreover, important discoveries would continue in solid-state physics. But none of these advances would bring about revolutionary changes in the foundations of physics, according to Bethe. As an example, Bethe cited the discovery of so-called high-temperature superconductors, arguably the most exciting advance in physics in decades. These materials, first reported in 1987, conduct electricity with no resistance at relatively high temperatures (which are still well below zero degrees Celsius). "That didn't in any way change our understanding of electric conduction or even of superconductivity," Bethe said. "The basic structure of quantum mechanics, quantum mechanics without relativity, that basic structure is finished." In fact, "the understanding of atoms, molecules, the chemical bond, and so on, that was all complete by 1928." Could there ever be another revolution in physics like the one that accompanied quantum mechanics? "That's very unlikely," Bethe replied in his unsettlingly matter-of-fact way.

Actually, almost all believers in a final theory agree that whatever form i takes, it will still be a *quantum* theory. Steven Weinberg suggested to me that a final theory of physics "might be as far removed from what we now understand as quantum mechanics was from classical mechanics." But he, like Hans Bethe, did not think the final theory would supplant quantum mechanics in any way. "I think we'll be stuck with quantum mechanics," Weinberg said. "So in that sense the development of quantum mechanics may be more revolutionary than anything before or after."

Weinberg's remarks reminded me of an essay published in *Physics Today* in 1990, in which the physicist David Mermin of Cornell University recounted how a certain Professor Mozart (actually Mermin's cranky alter ego) had complained that "particle physics over the last 40 or 50 years has been a disappointment. Who would have expected that in half a century we wouldn't learn anything really profound!" When Mermin asked the fictional professor what he meant, he replied, "All particle physics has taught us about the central mystery is that quantum mechanics still works. Perfectly, as far as anybody can tell. What a letdown!"[16]

John Wheeler and the "It from Bit"

Bethe, Weinberg, and Mermin all seemed to suggest that quantum mechanics is—at least in a qualitative sense—the final theory of physics. Some physicists and philosophers have proposed that if they could only understand quantum mechanics, if they could determine its meaning, they might find *The Answer.* One of the most influential and inventive interpreters of quantum mechanics, and of modern physics in general, is John Archibald Wheeler. Wheeler is the archetypal physics-for-poets physicist. He is famed for his analogies and aphorisms, self-made and coopted. Among the one-liners he bestowed on me when I interviewed him on a warm spring day at Princeton were "If I can't picture it, I can't understand it" (Einstein); "Unitarianism [Wheeler's nominal religion] is a feather bed to catch falling Christians" (Darwin); "Never run after a bus or woman or cosmological theory, because there'll always be another one in a few minutes" (a friend of Wheeler's at Yale); and "If you haven't found something strange during the day it hasn't been much of a day" (Wheeler).

Wheeler is also renowned for his physical energy. When we left his third-floor office to get some lunch, he spurned the elevator—"elevators are hazardous to your health," he declared—and charged down the stairs. He hooked an arm inside the banister and pivoted at each landing, letting centrifugal force whirl him around the hairpin and down the next flight. "We have contests to see who can take the stairs fastest," he said over a shoulder. Outside, Wheeler marched rather than walked, swinging his fists smartly in rhythm with his stride. He paused only when we reached a door. Invariably he got there first and yanked it open for me. After passing through I paused in reflexive deference—Wheeler was almost 80 at the time—but a moment later he was past me, barreling toward the next doorway.

The metaphor was so obvious I almost suspected Wheeler had it in mind. Wheeler had made a career of racing ahead of other scientists and throwing open doors for them. He had helped gain acceptance—or at least attention—for some of the most outlandish ideas of modern physics, from black holes to multiple-universe theories to quantum mechanics itself. Wheeler might have been dismissed as fun but flaky long ago if he had not had such unassailable credentials. In his early twenties, he had traveled to Denmark to study under Niels Bohr ("because he sees further than any man alive," Wheeler wrote in his application for the fellowship). Bohr was to be the most profound influence on Wheeler's thought. In 1939, Bohr and Wheeler published the first paper that successfully explained nuclear fission in terms of quantum physics.[17]

Wheeler's expertise in nuclear physics led to his involvement in the construction of the first fission-based bomb during World War II and, in the early years of the cold war, the first hydrogen bomb. After the war, Wheeler became one of the leading authorities on general relativity, Einstein's theory of gravity. He coined the term *black hole* in the late 1960s, and he played a major role in convincing astronomers that these bizarre, infinitely dense objects predicted by Einstein's theory might actually exist. I asked Wheeler what characteristic made him able to believe in such fantastical objects, which other physicists came to accept only with great reluctance. "More vividness of imagination," he replied. "There's that phrase of Bohr's I like so much: 'You must be prepared for a surprise, and a very great surprise.' "

Beginning in the 1950s, Wheeler had grown increasingly intrigued by the philosophical implications of quantum physics. The most widely accepted interpretation of quantum mechanics was the so-called orthodox interpretation (although "orthodox" seems an odd descriptor for such a radical worldview). Also called the Copenhagen interpretation, because it was set forth by Wheeler's mentor, Bohr, in a series of speeches in Copenhagen in the late 1920s, it held that subatomic entities such as electrons have no real existence; they exist in a probabilistic limbo of many possible superposed states until forced into a single state by the act of observation. The electrons or photons may act like waves or like particles, depending on how they are experimentally observed.

Wheeler was one of the first prominent physicists to propose that reality might not be wholly physical; in some sense, our cosmos might be a participatory phenomenon, requiring the act of observation—and thus consciousness itself. In the 1960s Wheeler helped to popularize the notorious anthropic principle. It holds, essentially, that the universe must be as it is, because, if it were otherwise, we might not be here to observe it. Wheeler also began to draw his colleagues' attention to some intriguing links between physics and information theory, which was invented in 1948 by the mathematician Claude Shannon. Just as physics builds on an elementary, indivisible entity—namely, the quantum—which is defined by the act of observation, so does information theory. Its quantum is the binary unit, or bit, which is a message representing one of two choices: heads or tails, yes or no, zero or one.

Wheeler became even more deeply convinced of the importance of information after concocting a thought experiment that exposed the strangeness of the quantum world for all to see. Wheeler's delayed-choice experiment is a variation on the classic (but not classical) two-slit experiment, which demonstrates the schizophrenic nature of quantum phenomena. When electrons are aimed at a barrier containing two slits, the electrons act like waves; they go through both slits at once and form what is called an interference pattern, created by the overlapping of the waves, when they strike a detector on the far side of the barrier. If the physicist closes off one slit at a time, however, the electrons pass through the open slit like simple particles and the interference pattern disappears. In the delayed-choice experiment, the experimenter decides whether to leave both slits open or to close one off *after the electrons have already passed through the barrier*—with the same results. The electrons seem to know in advance how the physicist will choose to observe them. This experiment was carried out in the early 1990s and confirmed Wheeler's prediction.

Wheeler accounted for this conundrum with yet another analogy. He likened the job of a physicist to that of someone playing 20 questions in its surprise version. In this variant of the old game, one person leaves the room while the rest of the group—or so the excluded person thinks— selects some person, place, or thing. The single player then reenters the room and tries to guess what the others have in mind by asking a series of questions that can only be answered yes or no. Unbeknownst to the guesser, the group has decided to play a trick. The first person to be queried will think of an object only *after* the questioner asks the question. Each person will do the same, giving a response that is consistent not only with the immediate question but also with all previous questions.

"The word wasn't in the room when I came in even though I thought it was," Wheeler explained. In the same way, the electron, before the physicist chooses how to observe it, is neither a wave nor a particle. It is in some sense unreal; it exists in an indeterminate limbo. "Not until you start asking a question, do you get something," Wheeler said. "The situation cannot declare itself until you've asked your question. But the asking of one question prevents and excludes the asking of another. So if you ask where my great white hope presently lies—and I always find it interesting to ask people what's your great white hope—I'd say it's in the idea that the whole show can be reduced to something similar in a broad sense to this game of 20 questions."

Wheeler has condensed these ideas into a phrase that resembles a Zen koan: "the it from bit." In one of his free-form essays, Wheeler unpacked the phrase as follows: ". . . every it—every particle, every field of force, even the spacetime continuum itself—derives its function, its meaning, its very existence entirely—even if in some contexts indirectly—from the apparatus-elicited answers to yes-or-no questions, binary choices, *bits*."[18]

Inspired by Wheeler, an ever-larger group of researchers—including computer scientists, astronomers, mathematicians, and biologists, as well as physicists—began probing the links between information theory and physics in the late 1980s. Some superstring theorists even joined in, trying to knit together quantum field theory, black holes, and information theory with a skein of strings. Wheeler acknowledged that these ideas were still raw, not yet ready for rigorous testing. He and his fellow explorers were still "trying to get the lay of the land" and "learning how to express things that we already know" in the language of information theory. The effort may lead to a dead end, Wheeler said, or to a powerful new vision of reality, of "the whole show."

Wheeler emphasized that science has many mysteries left to explain. "We live still in the childhood of mankind," he said. "All these horizons are beginning to light up in our day: molecular biology, DNA, cosmology. We're just children looking for answers." He served up another aphorism: "As the island of our knowledge grows, so does the shore of our ignorance." Yet he was also convinced that humans would someday find *The Answer*. In search of a quotation that expressed his faith, he jumped up and pulled down a book on information theory and physics to which he had contributed an essay. After flipping it open, he read: "Surely someday, we can believe, we will grasp the central idea of it all as so simple, so beautiful, so compelling that we will all say to each other, 'Oh, how could it have been otherwise! How could we all have been so blind for so long!' "[19] Wheeler looked up from the book; his expression was beatific. "I don't know whether it will be one year or a decade, but I think we can and will understand. That's the central thing I would like to stand for. We can and will understand."

Many modern scientists, Wheeler noted, shared his faith that humans would one day find *The Answer*. For example, Kurt Gödel, once Wheeler's neighbor in Princeton, believed that *The Answer* might *already* have been discovered. "He thought that maybe among the papers of Leibniz, which in his time had still not been fully smoked out, we would find the—what was the word—the philosopher's key, the magic way to find truth and solve any set of puzzlements." Gödel felt that this key "would give a person who understood it such power that you could only entrust the knowledge of this philosopher's key to people of high moral character."

Yet Wheeler's own mentor, Bohr, apparently doubted whether science or mathematics could achieve such a revelation. Wheeler learned of Bohr's view not from the great man himself, but from his son. After Bohr died, his son told Wheeler that Bohr had felt that the search for the ultimate theory of physics might never reach a satisfying conclusion; as physicists sought to penetrate further into nature, they would face questions of increasing complexity and difficulty that would eventually overwhelm them. "I guess I'm more optimistic than that," Wheeler said. He paused a moment and added, with a rare note of somberness, "But maybe I'm kidding myself."

The irony is that Wheeler's own ideas suggest that a final theory will always be a mirage, that the truth is in some sense imagined rather than objectively apprehended. According to the it from bit, we create not only truth, but even reality itself—the "it"—with the questions we ask. Wheeler's view comes dangerously close to relativism, or worse. In the early 1980s, organizers of the annual meeting of the American Association for the Advancement of Science placed Wheeler on a program with three parapsychologists. Wheeler was furious. At the meeting, he made it clear that he did not share the belief of his cospeakers in psychic phenomena. He passed out a pamphlet that declared, in reference to parapsychology, "Where there's smoke, there's smoke."

Yet Wheeler himself has suggested that there is nothing *but* smoke. "I do take 100 percent seriously the idea that the world is a figment of the imagination," he once remarked to the science writer and physicist Jeremy Bernstein.[20] Wheeler is well aware that this view is, from an empirical viewpoint, unsupportable: Where was mind when the universe was born? And what sustained the universe for the billions of years before we came to be? He nonetheless bravely offers us a lovely, chilling paradox: at the heart of everything is a question, not an answer. When we peer down into the deepest recesses of matter or at the farthest edge of the universe, we see, finally, our own puzzled faces looking back at us.

David Bohm's Implicate Order

Not surprisingly, some other physicist-philosophers have bridled both at Wheeler's views and, more generally, at the Copenhagen interpretation set forth by Bohr. One prominent dissident was David Bohm. Born and raised in Pennsylvania, Bohm left the United States in 1951, at the height of the McCarthy era, after refusing to answer questions from the Un-American Activities Committee about whether he or any of his scientific colleagues (notably Robert Oppenheimer) were communists. After stays in Brazil and Israel, he settled in England in the late 1950s.

By then, Bohm had already begun developing an alternative to the Copenhagen interpretation. Sometimes called the pilot-wave interpretation, it preserves all the predictive power of quantum mechanics but eliminates many of the most bizarre aspects of the orthodox interpretation, such as the schizophrenic character of quanta and their dependence on observers for existence. Since the late 1980s, the pilot-wave theory has attracted increasing attention from physicists and philosophers uncomfortable with the subjectivism and antideterminism of the Copenhagen interpretation.

Paradoxically, Bohm also seemed intent on making physics even *more* philosophical, speculative, holistic. He went much further than Wheeler did in drawing analogies between quantum mechanics and Eastern religion. He developed a philosophy, called the implicate order, that sought to embrace both mystical and scientific knowledge. Bohm's writings on these topics attracted an almost cultlike following; he became a hero to those who hoped to achieve mystical insight through physics. Few scientists combine these two contradictory impulses—the need to clarify reality and to mystify it—in such a dramatic fashion.[21]

In August 1992, I visited Bohm at his home in Edgeware, a suburb north of London. His skin was alarmingly pale, especially in contrast to his purplish lips and dark, wiry hair. His frame, sinking into a large armchair, seemed limp and langorous, but at the same time suffused with nervous energy. He cupped one hand over the top of his head; the other rested on the armrest. His fingers, long and blue veined, with tapered, yellow nails, were splayed. He was recovering, he told me, from a recent heart attack.

Bohm's wife brought us tea and biscuits and then retreated to another part of the house. Bohm spoke haltingly at first, but gradually the words came faster, in a low, urgent monotone. His mouth was apparently dry, because he kept smacking his lips. Occasionally, after making some observation that amused him, he pulled his lips back from his teeth in a semblance of a smile. He also had the disconcerting habit of pausing every few sentences and saying, "Is that clear?" or simply, "Hmmm?" I was often so befuddled, and so hopeless of finding my way to understanding, that I just nodded and smiled. But Bohm could be absolutely, rivetingly clear, too. Later, I learned that he had this same effect on others; like some strange quantum particle, he oscillated in and out of focus.

Bohm said he had begun to question the Copenhagen interpretation in the late 1940s, while writing a book on quantum mechanics. Bohr had rejected the possibility that the probabilistic behavior of quantum systems was actually the result of underlying, deterministic mechanisms, sometimes called hidden variables. Reality was unknowable because it was intrinsically indeterminate, Bohr insisted.

Bohm found this view unacceptable. "The whole idea of science so far has been to say that underlying the phenomenon is some reality which explains things," he explained. "It was not that Bohr denied reality, but he said quantum mechanics implied there was nothing more that could be said about it." Such a view, Bohm decided, reduced quantum mechanics to "a system of formulas that we use to make predictions or to control things technologically. I said, that's not enough. I don't think I would be very interested in science if that were all there was."

In a paper published in 1952, Bohm proposed that particles are indeed particles—and at all times, not just when they are observed. Their behavior is determined by a new, heretofore undetected force, which Bohm called the pilot wave. Any effort to measure these properties precisely would destroy information about them by physically altering the pilot wave. Bohm thus gave the uncertainty principle a purely physical rather than a metaphysical meaning. Bohr had interpreted the uncertainty principle as meaning "not that there is uncertainty, but that there is an inherent ambiguity" in a quantum system, Bohm told me.

Bohm's interpretation did permit, and even highlight, one quantum paradox: nonlocality, the ability of one particle to influence another instantaneously across vast distances. Einstein had drawn attention to nonlocality in 1935 in an effort to show that quantum mechanics must be flawed. Together with Boris Podolsky and Nathan Rosen, Einstein proposed a thought experiment—now called the EPR experiment—involving two particles that spring from a common source and fly in opposite directions.[22]

According to the standard model of quantum mechanics, neither particle has a definite position or momentum before it is measured; but by measuring the momentum of one particle, the physicist instantaneously forces the other particle to assume a fixed position—even if it is on the other side of the galaxy. Deriding this effect as "spooky action at a distance," Einstein argued that it violated both common sense and his own theory of special relativity, which prohibits the propagation of effects faster than the speed of light; quantum mechanics must therefore be an incomplete theory. In 1980, however, a group of French physicists carried out a version of the EPR experiment and showed that it did indeed give rise to spooky action. (The reason that the experiment does not violate special relativity is that one cannot exploit nonlocality to transmit information.)

Bohm never had any doubts about the outcome of the experiment. "It would have been a terrific surprise to find out otherwise," he said.

As Bohm was trying to bring the world into sharper focus through his pilot-wave model, however, he was also arguing that complete clarity was impossible. His ideas were inspired in part by an experiment he saw on television, in which a drop of ink was squeezed onto a cylinder of glycerin. When the cylinder was rotated, the ink diffused through the glycerin in an apparently irreversible fashion; its order seemed to have disintegrated. But when the direction of rotation was reversed, the ink gathered into a drop again.

Upon this simple experiment, Bohm built a worldview called the implicate order. Underlying the apparently chaotic realm of physical appearances—the explicate order—there is always a deeper, hidden, implicate order. Applying this concept to the quantum realm, Bohm proposed that the implicate order is the quantum potential, a field consisting of an infinite number of fluctuating pilot waves. The overlapping of these waves generates what appear to us as particles, which constitute the explicate order. Even such seemingly fundamental concepts as space and time may be merely explicate manifestations of some deeper, implicate order, according to Bohm.

To plumb the implicate order, Bohm said, physicists might need to jettison some basic assumptions about the organization of nature. "Fundamental notions like order and structure condition our thinking unconsciously, and new kinds of theories depend on new kinds of order," he remarked. During the Enlightenment, thinkers such as Newton and Descartes replaced the ancients' organic concept of order with a mechanistic view. Although the advent of relativity and other theories brought about modifications in this order, "the basic idea is still the same," Bohm said, "a mechanical order described by coordinates."

But Bohm, in spite of his own enormous ambitions as a truth seeker, rejected the possibility that scientists could bring their enterprise to an end by reducing all the phenomena of nature to a single phenomenon (such as superstrings). "I think there are no limits to this. People are going to talk about the theory of everything, but that's an assumption, you see, which has no basis. At each level we have something which is taken as appearance and something else is taken as the essence which explains the appearance. But then when we move to another level essence and appearance interchange their rules, right? Is that clear? There's no end to this, you see? The very nature of our knowledge is of that nature, you see. But what underlies it all is unknown and cannot be grasped by thought."

To Bohm, science was "an inexhaustible process." Modern physicists, he pointed out, assume that the forces of nature are the essence of reality. "But why are the forces of nature there? The forces of nature are then taken as the essence. The atoms weren't the essence. Why should these forces be?"

The belief of modern physicists in a final theory could only be self-fulfilling, Bohm said. "If you do that you'll keep away from really questioning deeply." He noted that "if you have fish in a tank and you put a glass barrier in there, the fish keep away from it. And then if you take away the glass barrier they never cross the barrier and they think the whole world is that." He chuckled drily. "So your thought that this is the end could be the barrier to looking further."

Bohm reiterated that "we're not ever going to get a final essence which isn't also the appearance of something." But wasn't that frustrating? I asked. "Well, it depends what you want. You're frustrated if you want to get it all. On the other hand, scientists are going to be frustrated if they get the final answer and then have nothing to do except be technicians, you see." He uttered his dry laugh. So, I said, damned if you do and damned if you don't. "Well, I think you have to look at it differently, you see. One of the reasons for doing science is the extension of perception and not of current knowledge. We are constantly coming into contact with reality better and better."

Science, Bohm continued, is sure to evolve in totally unexpected ways. He expressed the hope that future scientists would be less dependent on mathematics for modeling reality and would draw on new sources of metaphor and analogy. "We have an assumption now that's getting stronger and stronger that mathematics is the only way to deal with reality," Bohm said. "Because it's worked so well for a while we've assumed that it has to be that way."

Like many other scientific visionaries, Bohm expected that science and art would someday merge. "This division of art and science is temporary," he observed. "It didn't exist in the past, and there's no reason why it should go on in the future." Just as art consists not simply of works of art but of an "attitude, the artistic spirit," so does science consist not in the accumulation of knowledge but in the creation of fresh modes of perception. "The ability to perceive or think differently is more important than the knowledge gained," Bohm explained. There was something poignant about Bohm's hope that science might become more artlike. Most physicists have objected to his pilot-wave interpretation on aesthetic grounds: it is too ugly to be right.

Bohm, trying to convince me once and for all of the impossibility of final knowledge, offered the following argument: "Anything known has to be determined by its limits. And that's not just quantitative but qualitative. The theory is this and not that. Now it's consistent to propose that there is the unlimited. You have to notice that if you say there is the unlimited, it cannot be different, because then the unlimited will limit the limited, by saying that the limited is not the unlimited, right? The unlimited must include the limited. We have to say, from the unlimited the limited arises, in a creative process; that's consistent. Therefore we say that no matter how far we go there is the unlimited. It seems that no matter how far you go, somebody will come up with another point you have to answer. And I don't see how you could ever settle that."

At this moment, to my relief, Bohm's wife entered the room and asked if we wanted more tea. As she refilled my cup, I pointed out a book on Tibetan mysticism on a shelf behind Bohm. When I asked Bohm if he had been influenced by such writings, he nodded. He had been a friend and student of the Indian mystic Krishnamurti, who died in 1986. Krishnamurti was one of the first modern Indian sages to try to show westerners how to achieve enlightenment. Was Krishnamurti himself enlightened? "In some ways, yes," Bohm replied. "His basic thing was to go into thought, to get to the end of it, completely, and thought would become a different kind of consciousness." Of course, one could never truly plumb one's own mind, Bohm said. Any attempt to examine one's own thought changes it—just as the measurement of an electron alters its course. There could be no final mystical knowledge, Bohm seemed to be implying, any more than there could be a final theory of physics.

Was Krishnamurti a happy person? Bohm looked puzzled at my question. "That's hard to say," he replied eventually. "He was unhappy at times, but I think he was pretty happy overall. The thing is not about happiness, really." Bohm frowned, as if realizing the import of what he had just said.

In *Science, Order, and Creativity*, cowritten with F. David Peat, Bohm insisted on the importance of "playfulness" in science, and in life.[23] But

Bohm himself, both in his writings and in person, was anything but playful. For him, this was not a game, this truth seeking; it was a dreadful, impossible, but necessary task. Bohm was desperate to know, to discover the secret of everything, either through physics or through meditation, through mystical knowledge. And yet he insisted that reality was unknowable—because, I believe, he was repelled by the thought of finality. He recognized that any truth, no matter how initially wondrous, eventually ossifies into a dead, inanimate thing that does not reveal the absolute but conceals it. Bohm craved not truth but revelation, perpetual revelation. As a result, he was doomed to perpetual doubt.

I finally said good-bye to Bohm and his wife and departed. Outside, a light rain was falling. I walked up the path to the street and glanced back at the Bohms' house, a modest, whitewashed cottage on a street of modest whitewashed cottages. Bohm died of a heart attack two months later.[24]

Feynman's Gloomy Prophecy

In *The Character of Physical Law*, Richard Feynman, who won a Nobel Prize in 1965 for devising a quantum version of electromagnetism, offered a rather dark prophecy about the future of physics:

> We are very lucky to live in an age in which we are still making discoveries. It is like the discovery of America—you only discover it once. The age in which we live is the age in which we are discovering the fundamental laws of nature, and that day will never come again. It is very exciting, it is marvelous, but this excitement will have to go. Of course in the future there will be other interests. There will be the interest of the connection of one level of phenomena to another— phenomena in biology and so on, or, if you are talking about exploration, exploring other planets, but there will not be the same things we are doing now.[25]

After the fundamental laws are discovered, physics will succumb to second-rate thinkers, that is, philosophers. "The philosophers who are always on the outside making stupid remarks will be able to close in, because we cannot push them away by saying, 'If you were right we would be able to guess all the laws,' because when the laws are all there they will have an explanation for them. . . . There will be a degeneration of ideas, just like the degeneration that great explorers feel is occurring when tourists begin moving in on a new territory."[26]

Feynman's vision was uncannily on target. He erred only in thinking that it would be millennia, not decades, before the philosophers closed in. I saw the future of physics in 1992 when I attended a symposium at Columbia University in which philosophers and physicists discussed the meaning of quantum mechanics.[27] The symposium demonstrated that more than 60 years after quantum mechanics was invented, its meaning remained, to put it politely, elusive. In the lectures, one could hear echoes of Wheeler's it from bit approach, and Bohm's pilot-wave hypothesis, and the many-worlds model favored by Steven Weinberg and others. But for the most part each speaker seemed to have arrived at a private understanding of quantum mechanics, couched in idiosyncratic language; no one seemed to understand, let alone agree with, anyone else. The bickering brought to mind what Bohr once said of quantum mechanics: "If you think you understand it, that only shows you don't know the first thing about it."[28]

Of course, the apparent disarray could have stemmed entirely from my own ignorance. But when I revealed my impression of confusion and dissonance to one of the attendees, he reassured me that my perception was accurate. "It's a mess," he said of the conference (and, by implication, the whole business of interpreting quantum mechanics). The problem, he noted, arose because, for the most part, the different interpretations of quantum mechanics cannot be empirically distinguished from one another; philosophers and physicists favor one interpretation over another for aesthetic and philosophical—that is, subjective—reasons.

This is the fate of physics. The vast majority of physicists, those employed in industry and even academia, will continue to apply the knowledge they already have in hand—inventing more versatile lasers and superconductors and computing devices—without worrying about any underlying philosophical issues.[29] A few diehards dedicated to truth rather than practicality will practice physics in a nonempirical, ironic mode, plumbing the magical realm of superstrings and other esoterica and fretting about the meaning of quantum mechanics. The conferences of these ironic physicists, whose disputes cannot be experimentally resolved, will become more and more like those of that bastion of literary criticism, the Modern Language Association.

CHAPTER FOUR

The End of Cosmology

In 1990 I traveled to a remote resort in the mountains of northern Sweden to attend a symposium entitled "The Birth and Early Evolution of Our Universe." When I arrived, I found that about 30 particle physicists and astronomers from around the world—the United States, Europe, the Soviet Union, and Japan—were there. I had come to the meeting in part to meet Stephen Hawking. The compelling symbolism of his plight—powerful brain in a paralyzed body—had helped to make him one of the best-known scientists in the world.

Hawking's condition, when I met him, was worse than I had expected. He sat in a semifetal position, hunch shouldered, slack jawed, and painfully frail, his head tipped to one side, in a wheelchair loaded with batteries and computers. As far as I could tell, he could move only his left forefinger. With it he laboriously selected letters, words, or sentences from a menu on his computer screen. A voice synthesizer uttered the words in an incongruously deep, authoritative voice—reminiscent of the cyborg hero of *Robocop*. Hawking seemed, for the most part, more amused than distressed by his plight. His purple-lipped, Mick Jagger mouth often curled up at one corner in a kind of smirk.

Hawking was scheduled to give a talk on quantum cosmology, a field he had helped to create. Quantum cosmology assumes that at very small scales, quantum uncertainty causes not merely matter and energy, but the very fabric of space and time, to flicker between different states. These space-time fluctuations might give rise to wormholes, which could link one region of space-time with another one very far away, or to "baby universes." Hawking had stored the hour-long speech, titled "The Alpha

Parameters of Wormholes," in his computer; he had merely to tap a key to prompt his voice synthesizer to read it, sentence by sentence.

In his eerie cyber-voice, Hawking discussed whether we might someday be able to slip into a wormhole in our galaxy and a moment later pop out the other end in a galaxy far, far away. Probably not, he concluded, because quantum effects would scramble our constituent particles beyond recognition. (Hawking's argument implied that "warp drive," the faster-than-light transport depicted in *Star Trek*, was impossible.) He wrapped up his lecture with a digression on superstring theory. Although all we see around us is the "mini-superspace" that we call space-time, "we are really living in the infinite dimensional superspace of string theory."[1]

My response to Hawking was ambivalent. He was, on the one hand, a heroic figure. Trapped in a crippled, helpless body, he could still imagine realities with infinite degrees of freedom. On the other hand, what he was saying struck me as being utterly preposterous. Wormholes? Baby universes? Infinite dimensional superspace of string theory? This seemed more science fiction than science.

I had more or less the same reaction to the entire conference. A few talks—those in which astronomers discussed what they had gleaned from probing the cosmos with telescopes and other instruments—were firmly grounded in reality. This was empirical science. But many of the presentations addressed issues hopelessly divorced from reality, from any possible empirical test. What was the universe like when it was the size of a basketball, or a pea, or a proton, or a superstring? What is the effect on our universe of all the other universes linked to it by wormholes? There was something both grand and ludicrous about grown men (no women were present) bickering over such issues.

Over the course of the meeting, I struggled to quell that instinctive feeling of preposterousness, with some success. I reminded myself that these were terribly smart people, "the greatest geniuses in the world," as a local Swedish newspaper had put it. They would not waste their time on trivial pursuits. I therefore did my best, in writing later about the ideas of Hawking and other cosmologists, to make them sound plausible, to instill awe and comprehension instead of skepticism and confusion in readers. That is the job of the science writer, after all.

But sometimes the clearest science writing is the most dishonest. My initial reaction to Hawking and others at the conference was, to some extent, appropriate. Much of modern cosmology, particularly those aspects inspired by unified theories of particle physics and other esoteric ideas, *is* preposterous. Or, rather, it is ironic science, science that is not experimentally testable or resolvable even in principle and therefore is not science in the strict sense at all. Its primary function is to keep us awestruck before the mystery of the cosmos.

The irony is that Hawking was the first prominent physicist of his generation to predict that physics might soon achieve a complete, unified theory of nature and thus bring about its own demise. He offered up this prophecy in 1980, just after he had been named the Lucasian Professor of Mathematics at the University of Cambridge; Newton had held this chair some 300 years earlier. (Few observers noted that at the end of his speech, titled "Is the End of Theoretical Physics in Sight?", Hawking suggested that computers, given their accelerated evolution, might soon surpass their human creators in intelligence and achieve the final theory on their own.)[2] Hawking spelled out his prophecy in more detail in *A Brief History of Time*. The attainment of a final theory, he declared in the book's closing sentence, might help us to "know the mind of God."[3] The phrase suggested that a final theory would bequeath us a mystical revelation in whose glow we could bask for the rest of time.

But earlier in the book, in discussing what he called the no-boundary proposal, Hawking offered a very different view of what a final theory might accomplish. The no-boundary proposal addressed the age-old questions: What was there before the big bang? What exists beyond the borders of our universe? According to the no-boundary proposal, the entire history of the universe, all of space and all of time, forms a kind of four-dimensional sphere: space-time. Talking about the beginning or end of the universe is thus as meaningless as talking about the beginning or end of a sphere. Physics, too, Hawking conjectured, might form a perfect, seamless whole after it is unified; there might be only one fully consistent unified theory capable of generating space-time as we know it. God might not have had any choice in creating the universe.

"What place, then, for a creator?" Hawking asked.[4] There is *no* place, was his reply; a final theory would exclude God from the universe, and with him all mystery. Like Steven Weinberg, Hawking hoped to rout mysticism, vitalism, creationism from one of their last refuges, the origin of the universe. According to one biographer, Hawking and his wife, Jane, separated in 1990 in part because she, as a devout Christian, had become increasingly offended by his atheism.[5]

After the publication of *Brief History*, several other books took up the question of whether physics could achieve a complete and final theory, one that would answer all questions and thereby bring physics to an end. Those who claimed that no such theory was possible tended to fall back on Gödel's theorem and other esoterica. But throughout his career, Hawking has demonstrated that there is a much more basic obstacle to a theory of everything. Physicists can never eradicate mystery from the universe, they can never find *The Answer*, as long as there are physicists with imaginations as florid as Hawking's.

I suspect that Hawking—who may be less a truth seeker than an artist, an illusionist, a cosmic joker—knew all along that finding and empirically validating a unified theory would be extremely difficult, even impossible. His declaration that physics was on the verge of finding *The Answer* may well have been an ironic statement, less an assertion than a provocation. In 1994, he admitted as much when he told an interviewer that physics might never achieve a final theory after all.[6] Hawking is a master practitioner of ironic physics and cosmology.

Cosmology's Great Surprises

The most remarkable fact about modern cosmology is that it is not *all* ironic. Cosmology has given us several genuine, irrefutable surprises. Early in this century, the Milky Way, the island of stars within which our own sun nestles, was thought to constitute the entire universe. Then astronomers realized that minute smudges of light called nebulae, thought to be mere clouds of gas within the Milky Way, were actually islands of stars. The Milky Way was only one of a vast number of galaxies in a universe that was much, much larger than anyone had imagined. That finding represented an enormous, empirical, irrevocable surprise, one that even the most hardcore relativist would be hard-pressed to deny. To paraphrase Sheldon Glashow, galaxies are not imagined and unimagined; they exist.

Another great surprise was to come. Astronomers discovered that the glow of galaxies was invariably shifted toward the red end of the visible spectrum. Apparently the galaxies were hurtling away from the earth and from each other, and this recessional velocity caused the light to undergo a

Doppler shift (the same shift that makes an ambulance siren deepen in pitch as it races away from a listener). The red-shift evidence supported a theory, based on Einstein's theory of relativity, that the universe had begun in an explosion that was ongoing.

In the 1950s, theorists predicted that the fiery birth of the universe billions of years ago should have left an afterglow in the form of faint microwaves. In 1964 two radio engineers at Bell Laboratories stumbled upon the so-called cosmic background radiation. Physicists had also proposed that the fireball of creation would have served as a nuclear furnace in which hydrogen fused into helium and other light elements. Careful observations over the past few decades have shown that the abundances of light elements in the Milky Way and in other galaxies precisely match theoretical predictions.

David Schramm of Fermilab and the University of Chicago likes to call these three lines of evidence—the red shift of galaxies, the microwave background, and the abundance of elements—the pillars on which the big bang theory stands. Schramm is a big, barrel-chested, ebullient man, a pilot, mountain climber, and former collegiate champion in Greco-Roman wrestling. He is an indefatigable booster of the big bang—and of his own role in refining the calculations of light-element abundances. After I arrived at the symposium in Sweden, Schramm sat me down and went over the evidence for the big bang in great detail. "The big bang is in *fantastic* shape," he said. "We have the basic framework. We just need to fill in the gaps."

Schramm acknowledged that some of those gaps were rather large. Theorists cannot determine precisely how the hot plasma of the early universe condensed into stars and galaxies. Observations have suggested that the visible, starry stuff that astronomers can see through their telescopes is not massive enough to keep galaxies from flying apart; some invisible, or dark, matter must be binding the galaxies together. All the matter that we can see, in other words, may be just foam on a deep, dark sea.

Another question concerns what cosmologists like to call "large-scale structure." In the early days of cosmology, galaxies seemed to be scattered more or less evenly throughout the universe. But as observations improved, astronomers found that galaxies tend to huddle together in clusters surrounded by gigantic voids. Finally, there is the question of how the universe behaved in the so-called quantum gravity era, when the cosmos was so small and hot that all the forces of the universe were thought to be unified. These were the issues that dominated discussion during the Nobel symposium in Sweden. But none of these problems, Schramm insisted, threatened the basic framework of the big bang. "Just because you can't predict tornadoes," he said, "doesn't mean the earth isn't round."[7]

Schramm delivered much the same message to his fellow cosmologists at the Nobel symposium. He kept proclaiming that cosmology was in a "golden age." His chamber of commerce enthusiasm seemed to grate on some of his colleagues; after all, one does not become a cosmologist to fill in the details left by the pioneers. After Schramm's umpteenth "golden age" proclamation, one physicist snapped that you cannot know an age is golden when you are *in* that age but only in retrospect. Schramm jokes proliferated. One colleague speculated that the stocky physicist might represent the solution to the dark matter problem. Another proposed that Schramm be employed as a plug to prevent our universe from being sucked down a wormhole.

Toward the end of the meeting in Sweden, Hawking, Schramm, and all the other cosmologists piled into a bus and drove to a nearby village to hear a concert. When they entered the Lutheran church where the concert was to be held, it was already filled. The orchestra, a motley assortment of flaxen-haired youths and wizened elders clutching violins, clarinets, and other instruments, was already seated at the front of the church. Their neighbors jammed the balconies and the seats at the rear of the church. As the scientists filed down the center aisle to the pews at the front reserved for them, Hawking leading the way in his motorized wheelchair, the townspeople started to clap, tentatively at first, then passionately, for almost a full minute. The symbolism was perfect: for this moment, at least, in this place and for these people, science had usurped the role of religion as the source of truth about the universe.

Doubts had infiltrated the scientific priesthood, however. In the moments before the concert began, I overheard a conversation between David Schramm and Neil Turok, a young British physicist. Turok confided to Schramm that he was so concerned about the intractability of questions related to dark matter and the distribution of galaxies that he was thinking of quitting cosmology and entering another field. "Who says we have any right to understand the universe, anyway?" Turok asked plaintively.

Schramm shook his big head. The basic framework of cosmology, the big bang theory, was absolutely sound, he whispered insistently, as the orchestra began warming up; cosmologists just needed to tie up a few loose ends. "Things will sort themselves out," Schramm said.

Turok seemed to find Schramm's words comforting, but he probably should have been alarmed. What if Schramm was right? What if cosmologists already had, in the big bang theory, the major answer to the puzzle of the universe? What if all that remained was tying up loose ends, those that could be tied up? Given this possibility, it is no wonder that "strong" scientists such as Hawking have vaulted past the big bang theory into postempirical science. What else can someone so creative and ambitious do?

The Russian Magician

One of Stephen Hawking's few rivals as a practitioner of ironic cosmology is Andrei Linde, a Russian physicist who emigrated to Switzerland in 1988 and to the United States two years later. Linde, too, was at the Nobel symposium in Sweden, and his antics were among the highlights of the meeting. After imbibing a drink or two at an outdoor cocktail party, Linde snapped a rock in half with a karate chop. He stood on his hands and then flipped himself backward and landed on his feet. He pulled a box of wooden matches out of his pocket and placed two of them, forming a cross, on his hand. While Linde kept his hand—at least seemingly—perfectly still, the top match trembled and hopped as if jerked by an invisible string. The trick maddened his colleagues. Before long, matches and curses were flying every which way as a dozen or so of the world's most prominent cosmologists sought in vain to duplicate Linde's feat. When they demanded to know how Linde did it, he smiled and growled, "Ees kvantum fluctuation."

Linde is even more renowned for his theoretical sleights of hand. In the early 1980s he helped to win acceptance for one of the more extravagant ideas to emerge from particle physics: inflation. The invention of inflation (the term *discovery* is not appropriate here) is generally credited to Alan Guth of MIT, but Linde helped to refine the theory and win its acceptance. Guth and Linde proposed that very early in the history of our universe—at $T = 10^{-43}$ seconds, to be precise, when the cosmos was allegedly much smaller than a proton—gravity might briefly have become a repulsive rather than an attractive force. As a result, the universe supposedly passed through a tremendous, exponential growth spurt before settling down to its current, much more leisurely rate of expansion.

Guth and Linde based their idea on untested—and almost certainly untestable—unified theories of particle physics. Cosmologists nonetheless fell in love with inflation, because it could explain some nagging problems raised by the standard big bang model. First, why does the universe appear more or less the same in all directions? The answer is that just as blowing up a balloon smooths out its wrinkles, so would the exponential expansion of the universe render it relatively smooth. Conversely, inflation also explains why the universe is not a *completely* homogeneous consommé of radiation, but contains lumps of matter in the form of stars and galaxies. Quantum mechanics suggests that even empty space is brimming with energy; this energy constantly fluctuates, like waves dancing on the surface of a windblown lake. According to inflation, the peaks generated by these quantum fluctuations in the very early universe could have become large enough, after being inflated, to serve as the gravitational seeds from which stars and galaxies would grow.

Inflation has some startling implications, one of which is that everything we can see through our telescopes represents an infinitesimal fraction of the vastly larger realm created during inflation. But Linde did not stop there. Even that prodigious universe, he contended, is only one of an infinite number spawned by inflation. Inflation, once it begins, can never end; it has created not only our universe—the galaxy-emblazoned realm we ponder through our telescopes—but also countless others. This mega-verse has what is known as a fractal structure: the big universes sprout little universes, which sprout still smaller ones, and so on. Linde called his model the chaotic, fractal, eternally self-reproducing, inflationary universe.[8]

For someone so publicly playful and inventive, Linde can be surprisingly dour. I glimpsed this side of his character when I visited him at Stanford University, where he and his wife, Renata Kallosh, who is also a theoretical physicist, began working in 1990. When I arrived at the gray, cubist home they were renting, Linde gave me a perfunctory tour. In the backyard we encountered Kallosh, who was rooting happily in a flower bed. "Look, Andrei!" she cried, pointing to a nest filled with cheeping birds on a tree branch above her. Linde, his pallor and squint betraying his unfamiliarity with sunlight, merely nodded. When I asked if he found California relaxing, he muttered, "Maybe too relaxing."

As Linde recounted his life story, it became apparent that anxiety, even depression, had played a significant role in motivating him. At various stages of his career, he would despair of having any insight into the nature of things—just before achieving a breakthrough. Linde had stumbled on the basic concept of inflation in the late 1970s, while in Moscow, but decided that the idea was too flawed to pursue. His interest was revived by Alan Guth's proposal that inflation could explain several puzzling features of the universe, such as its smoothness, but Guth's version, too, was flawed. After pondering the problem so obsessively that he developed an ulcer, Linde showed how Guth's model could be adjusted to eliminate the technical problems.

But even this model of inflation depended on features of unified theories that were, Linde felt, suspect. Eventually—after falling into a gloom so deep that he had difficulty getting out of bed—he determined that inflation could result from much more generic quantum processes first proposed by John Wheeler. According to Wheeler, if one had a microscope trillions upon trillions of times more powerful than any in existence, one would see space and time fluctuating wildly because of quantum uncertainty. Linde contended that what Wheeler called "space-time foam" would inevitably give rise to the conditions necessary for inflation.

Inflation is a self-exhausting process; the expansion of space quickly causes the energy driving inflation to dissipate. But Linde argued that once inflation begins, it will always continue somewhere—again, because of quantum uncertainty. (A handy feature, this quantum uncertainty.) New universes are sprouting into existence at this very moment. Some immediately collapse back into themselves. Others inflate so fast that matter never has a chance to coalesce. And some, like ours, settle down to an expansion leisurely enough for gravity to mold matter into galaxies, stars, and planets.

Linde sometimes compared this supercosmos to an infinite sea. Viewed up close, the sea conveys an impression of dynamism and change, of waves heaving up and down. We humans, because we live within one of these heaving waves, think that the entire universe is expanding. But if we could rise above the sea's surface, we would realize that our expanding cosmos is just a tiny, insignificant, local feature of an infinite, eternal ocean. In a way, Linde contended, the old steady-state theory of Fred Hoyle (which I discuss later in this chapter) was right; when seen from a God-like perspective, the supercosmos exhibits a kind of equilibrium.

Linde was hardly the first physicist to posit the existence of other universes. But whereas most theorists treat other universes as mathematical abstractions, and slightly embarrassing ones at that, Linde delighted in speculating about their properties. Elaborating on his self-reproducing universe theory, for example, Linde borrowed from the language of genetics. Each universe created by inflation gives birth to still other "baby universes." Some of these offspring retain the "genes" of their predecessors and evolve into similar universes with similar laws of nature—and perhaps similar inhabitants. Invoking the anthropic principle, Linde proposed that some cosmic version of natural selection might favor the perpetuation of universes that are likely to produce intelligent life. "The fact that somewhere else there is life like ours is to me almost certain," he said. "But we can never know this."

Like Alan Guth and several other cosmologists, Linde liked to speculate on the feasibility of creating an inflationary universe in a laboratory. But only Linde asked: Why would one *want* to create another universe? What purpose would it serve? Once some cosmic engineer created a new universe, it would instantaneously dissociate itself from its parent at faster-than-light speeds, according to Linde's calculations. No further communication would be possible.

On the other hand, Linde surmised, perhaps the engineer could manipulate the seed of preinflationary stuff in such a way that it would evolve into a universe with particular dimensions, physical laws, and constants of nature. In that way, the engineer could impress a message of some sort onto the very structure of the new universe. In fact, Linde suggested, our own universe might have been created by beings in another universe, and physicists such as Linde, in their fumbling attempts to unravel the laws of nature, might actually be decoding a message from our cosmic parents.

Linde presented these ideas warily, watching my reaction. Only at the end, perhaps taking satisfaction in my gaping mouth, did he permit himself a little smile. His smile faded, however, when I wondered what the message imbedded in our universe might be. "It seems," he said wistfully, "that we are not quite grown up enough to know." Linde looked even more glum when I asked if he ever worried that all his work might be—I struggled to find the right word—bullshit.

"In my moments of depression I feel myself a complete idiot," he replied. "What I am playing with are some very primitive toys." He added that he tried not to become too attached to his own ideas. "Sometimes the models are quite strange, and if you take them too seriously then you are in danger to be trapped. I would say this is similar to running over very thin ice on the lake. If you just are running very fast, you may not sink, and you may go a large distance. If you stay and think if you are running in the right direction, you may go down."

Linde seemed to be saying that his goal as a physicist was not to achieve resolution, to arrive at *The Answer* or even simply at *an* answer, but to keep moving, to keep skating. Linde feared the thought of finality. His self-reproducing universe theory makes sense in this light: if the universe is infinite and eternal, then so is science, the quest for knowledge. But even a physics confined to this universe, Linde suggested, cannot be close to resolution. "For example, you do not include consciousness. Physics studies matter, and consciousness is not matter." Linde agreed with John Wheeler that reality must be in some sense a participatory phenomenon. "Before you make measurement, there is no universe, nothing you can call objective reality," Linde said.

Linde, like Wheeler and David Bohm, seemed to be tormented by mystical yearnings that physics alone could never satisfy. "There is some limit to rational knowledge," he said. "One way to study the irrational is to jump into it and just meditate. The other is to study the boundaries of the irrational with the tools of rationality." Linde chose the latter route, because physics offered a way "not to say total nonsense" about the workings of the world. But sometimes, he confessed, "I'm depressed when I think I will die like a physicist."

The Deflation of Inflation

The fact that Linde has earned so much respect—he was courted by several U.S. universities before choosing Stanford—bears witness both to his rhetorical skills and to cosmologists' hunger for new ideas. Nonetheless, by the early 1990s, inflation and many of the other exotic ideas that had emerged from particle physics in the previous decade had begun losing support from mainstream cosmologists. Even David Schramm, who had been quite bullish on inflation when I met him in Sweden, had his doubts when I spoke to him several years later. "I *like* inflation," Schramm said, but it can never be thoroughly verified because it does not generate any unique predictions, predictions that cannot be explained in some other way. "You won't see that for inflation," Schramm continued, "whereas for the big bang itself you do see that. The beautiful, cosmic microwave background and the light-element abundances tell you, 'This is it.' There's no other way of getting these observations."

Schramm acknowledged that as cosmologists venture further back toward the beginning of time, their theories become more speculative. Cosmology needs a unified theory of particle physics to describe processes in the very early universe, but validating a unified theory may be extremely difficult. "Even if somebody comes up with a really beautiful theory, like superstring theory, there's not any way it can be tested. So you're not really doing the scientific method, where you make predictions and then check it. There's not that experimental check going on. It's more just mathematical consistency."

Could the field end up being like the interpretation of quantum mechanics, where the standards are primarily aesthetic? "That's a real problem I have with it," Schramm replied, "that unless one comes up with tests, we are into the more philosophical rather than physics area. The tests have to give the universe as we observe it, but that's more of a post-diction rather than a pre-diction." It is always possible that theoretical explorations of black holes, superstrings, Wheeler's it from bit, and other exotica might yield some sort of breakthough. "But until someone comes up with definitive tests," Schramm said, "or we're lucky enough to find a black hole that we can probe in a careful way, we're not going to have that kind of 'Eureka' where you're really confident you know the answer."

Realizing the significance of what he was saying, Schramm suddenly switched back to his customary boosterish, public relations mode. The fact that cosmologists are having so much difficulty advancing beyond the big bang model is a *good* sign, he insisted, falling back on an all-too-familiar argument. "For example, at the turn of the century, physicists were saying most of physics is solved. There are a few nagging little problems, but it's basically solved. And we found that was certainly not the case. In fact, what we find is, that usually is the clue that there's another big step coming. Just when you think the end is in sight, you find that's a wormhole into a whole new perspective into the universe. And I think that may be what's going to happen, that we're zooming in and starting to see things. We'll see certain nagging problems that we haven't been able to solve. And I expect the solution to those problems will lead to a whole new rich and exciting area. The enterprise is not going to die."[9]

But what if cosmology *has* passed its peak, in the sense that it is unlikely to deliver any more empirical surprises as profound as the big bang theory itself? Cosmologists are lucky to know anything with certainty, according to Howard Georgi, a particle physicist at Harvard University. "I think you have to regard cosmology as a historical science, like evolutionary biology," said Georgi, a cherub-faced man with a cheerfully sardonic manner. "You're trying to look at the present-day universe and extrapolate back, which is an interesting but dangerous thing to do, because there may have been accidents that had big effects. And they try very hard to understand what kinds of things can be accidental and what features are robust. But I find it difficult to understand those arguments well enough to really be convinced." Georgi suggested that cosmologists might acquire some much-needed humility by reading the books of the evolutionary biologist Stephen Jay Gould, who discusses the potential pitfalls of reconstructing the past based on our knowledge of the present (see Chapter 5).

Georgi chuckled, perhaps recognizing the improbability of any cosmologist's taking his advice. Like Sheldon Glashow, whose office was just down the hall, Georgi had once been a leader in the search for a unified theory of physics. And like Glashow, Georgi eventually denounced superstring theory and other candidates for a unified theory as nontestable and thus nonscientific. The fate of particle physics and cosmology, Georgi noted, are to some extent intertwined. Cosmologists hope that a unified theory will help them understand the universe's origins more clearly. Conversely, some particle physicists hope that in lieu of terrestrial experiments, they can find confirmation of their theories by peering through telescopes at the edge of the universe. "That strikes me as a bit of a push," Georgi remarked mildly, "but what can I say?" When I asked him about quantum cosmology, the field explored by Hawking, Linde, and others, Georgi smiled mischievously. "A simple particle physicist like myself has trouble in those uncharted waters," he said. He found papers on quantum cosmology, with all their talk of wormholes and time travel and baby universes, "quite amusing. It's like reading Genesis." As for inflation, it is "a wonderful sort of scientific myth, which is *at least as* good as any other creation myth I've ever heard."[10]

The Maverick of Mavericks

There will always be those who reject not only inflation, baby universes, and other highly speculative hypotheses, but the big bang theory itself. The dean of big bang bashers is Fred Hoyle, a British astronomer and physicist. A selective reading of Hoyle's resumé might make him appear the quintessential insider. He studied at the University of Cambridge under the Nobel laureate Paul Dirac, who correctly predicted the existence of antimatter. Hoyle became a lecturer at Cambridge in 1945, and in the 1950s he helped to show how stars forge the heavy elements of which planets and people are made. Hoyle founded the prestigious Institute of Astronomy at Cambridge in the early 1960s and served as its first director. For these and other achievements he was knighted in 1972. Yes, Hoyle is Sir Fred. Yet Hoyle's stubborn refusal to accept the big bang theory—and his adherence to fringe ideas in other fields—made him an outlaw in the field he had helped to create.[11]

Since 1988, Hoyle has lived in a high-rise apartment building in Bournemouth, a town on England's southern coast. When I visited him there, his wife, Barbara, let me in and took me into the living room, where I found Hoyle sitting in a chair watching a cricket match on television. He rose and shook my hand without taking his eyes off the match. His wife, gently admonishing him for his rudeness, went over to the TV and turned it off. Only then did Hoyle, as if waking from a spell, turn his full attention to me.

I expected Hoyle to be odd and embittered, but he was, for the most part, all too amiable. With his pug nose, jutting jaw, and penchant for slang—colleagues were "chaps" and a bogus theory a "bust flush"—he exuded a kind of blue-collar integrity and geniality. He seemed to revel in the role of outsider. "When I was young the old regarded me as an outrageous young fellow, and now that I'm old the young regard me as an outrageous old fellow." He chuckled. "I should say that nothing would embarrass me more than if I were to be viewed as someone who is repeating what he has been saying year after year," as many astronomers do. "What I would be worried about is somebody coming along and saying,

'What you've been saying is technically not sound.' That would worry me."
(Actually, Hoyle has been accused of both repetitiveness and technical
errors.)[12]

Hoyle had a knack for sounding reasonable—for example, when arguing that the seeds of life must have come to our planet from outer space.
The spontaneous generation of life on the earth, Hoyle once remarked,
would have been as likely as the assemblage of a 747 aircraft by a tornado
passing through a junkyard. Elaborating on this point during our interview, Hoyle pointed out that asteroid impacts rendered the earth uninhabi-
table until at least 3.8 billion years ago and that cellular life had almost
certainly appeared by 3.7 billion years ago. If one thought of the entire 4.5-
billion-year history of the planet as a 24-hour day, Hoyle elaborated, then
life appeared in about half an hour. "You've got to discover DNA; you've
got to make thousands of enzymes in that half an hour," he explained. "And
you've got to do it in a very hostile situation. So I find when you put all this
together it doesn't add up to a very attractive situation." As Hoyle spoke, I
found myself nodding in agreement. Yes, of course life could not have
originated here. What could be more obvious? Only later did I realize that
according to Hoyle's timetable, apes were transmogrified into humans
some 20 seconds ago, and modern civilization sprang into existence in less
than $\frac{1}{10}$ second. Improbable, perhaps, but it happened.

Hoyle had first started thinking seriously about the origin of the universe shortly after World War II, during long discussions with two other
physicists, Thomas Gold and Hermann Bondi. "Bondi had a relative
somewhere—he seemed to have relatives everywhere—and one sent him a
case of rum," Hoyle recalled. While imbibing Bondi's liquor, the three
physicists turned to a perennial puzzle of the young and intoxicated: How
did we come to be?

The finding that all galaxies in the cosmos are receding from one
another had already convinced many astronomers that the universe had
exploded into being at a specific time in the past and was still expanding.
Hoyle's fundamental objection to this model was philosophical. It did not
make sense to talk about the creation of the universe unless one already had
space and time for the universe to be created in. "You lose the universality
of the laws of physics," Hoyle explained to me. "Physics is no longer." The
only alternative to this absurdity, Hoyle decided, was that space and time
must have always existed. He, Gold, and Bondi thus invented the steady-

state theory, which posits that the universe is infinite both in space and time and constantly generates new matter through some still-unknown mechanism.

Hoyle stopped promoting the steady-state theory after the discovery of the microwave background radiation in the early 1960s seemed to provide conclusive evidence for the big bang. But his old doubts resurfaced in the 1980s, as he watched cosmologists struggle to explain the formation of galaxies and other puzzles. "I began to get the sense that there was something seriously wrong"—not only with new concepts such as inflation and dark matter, he said, but with the big bang itself. "I'm a great believer that if you have a correct theory, you show a lot of positive results. It seems to me that they'd gone on for 20 years, by 1985, and there wasn't much to show for it. And that couldn't be the case if it was right."

Hoyle thus resurrected the steady-state theory in a new and improved form. Rather than one big bang, he said, many little bangs occurred in preexisting space and time. These little bangs were responsible for light elements and the red shifts of galaxies. As for the cosmic microwave background, Hoyle's best guess was that it is radiation emitted by some sort of metallic interstellar dust. Hoyle acknowledged that his "quasi–steady state theory," which in effect replaced one big miracle with many little ones, was far from perfect. But he insisted that recent versions of the big bang theory, which posit the existence of inflation, dark matter, and other exotica, are much more deeply flawed. "It's like medieval theology," he exclaimed in a rare flash of anger.

The longer Hoyle spoke, however, the more I began to wonder just how sincere his doubts about the big bang were. In some of his statements, he revealed a proprietary fondness for the theory. One of the great ironies of modern science is that Hoyle coined the term *big bang* in 1950 while he was doing a series of radio lectures on astronomy. Hoyle told me that he meant not to disparage the theory, as many accounts have suggested, but merely to describe it. At the time, he recalled, astronomers often referred to the theory as "the Friedman cosmology," after a physicist who showed how Einstein's theory of relativity gave rise to an expanding universe.

"That was poison," Hoyle declared. "You had to have something vivid. So I thought up the big bang. If I had patented it, copyrighted it . . ." he mused. The August 1993 *Sky and Telescope* magazine initiated a contest to rename the theory. After mulling over thousands of suggestions, the judges announced they could find none worthy of supplanting *big bang*.[13] Hoyle said he was not surprised. "Words are like harpoons," he commented. "Once they go in, they are very hard to pull out."

Hoyle also seemed obsessed with how close he had come to discovering the cosmic microwave background. It was 1963, and during an astronomy conference Hoyle fell into a conversation with Robert Dicke, a physicist from Princeton who was planning to search for the cosmic microwaves predicted by the big bang model. Dicke told Hoyle that he expected the microwaves to be about 20 degrees above absolute zero, which is what most theorists were predicting. Hoyle informed Dicke that in 1941, a Canadian radio astronomer named Andrew McKellar had found interstellar gas radiating microwaves at 3 degrees, not 20.

To Hoyle's everlasting regret, neither he nor Dicke, during their conversation, spelled out the implication of McKellar's finding: that the microwave background might be 3 degrees. "We just sat there drinking coffee," Hoyle remembered, his voice rising. "If either of us had said, 'Maybe it *is* 3 degrees,' we'd have gone straightaway and checked it, and then we'd have had it in 1963." A year later, just before Dicke turned on his microwave experiment, Arno Penzias and Robert Wilson of Bell Laboratories discovered the 3-degree microwave radiation, an achievement for which they later received the Nobel Prize. "I've always felt that was one of the worst misses of my life," Hoyle sighed, shaking his head slowly.

Why should Hoyle care that he had nearly discovered a phenomenon that he now derided as spurious? I think Hoyle, like many mavericks, once hoped to be a member of the inner circle of science, draped in honor and glory. He went far toward realizing that goal. But in 1972, officials at Cambridge forced Hoyle to resign from his post as director of the Institute for Astronomy—for political rather than scientific reasons. Hoyle and his wife left Cambridge for a cottage on an isolated moor in northern England, where they lived for 15 years before moving to Bournemouth. During this period, Hoyle's anti-authoritarianism, which had always served him so well, became less creative than reactionary. He degenerated into what Harold Bloom derided as a "mere rebel," although he still dreamed of what might have been.

Hoyle also seemed to suffer from another problem. The task of the scientist is to find patterns in nature. There is always the danger that one will see patterns where there are none. Hoyle, in the latter part of his career, seemed to have succumbed to this pitfall. He saw patterns—or, rather, conspiracies—both in the structure of the cosmos and among those scientists who rejected his radical views. Hoyle's mind-set is most evident in his views on biology. Since the early 1970s he has argued that the universe is pervaded by viruses, bacteria, and other organisms. (Hoyle first broached this possibility in 1957 in *The Black Cloud*, which remains the best known of his many science fiction novels.) These space-faring microbes supposedly provided the seeds for life on earth and spurred evolution thereafter; natural selection played little or no role in creating the diversity of life.[14] Hoyle has also asserted that epidemics of influenza, whooping cough, and other diseases are triggered when the earth passes through clouds of pathogens.

Discussing the biomedical establishment's continued belief in the more conventional, person-to-person mode of disease transmission, Hoyle glowered. "They don't look at those data and say, 'Well, it's wrong,' and stop teaching it. They just go on doping out the same rubbish. And that's why if you go to the hospital and there's something wrong with you, you'll be lucky if they cure it." But if space is swarming with organisms, I asked, why haven't they been detected? Oh, but they probably were, Hoyle assured me. He suspected that U.S. experiments on high-altitude balloons and other platforms had turned up evidence of life in space in the 1960s, but that officials had hushed it up. Why? Perhaps for reasons related to national security, Hoyle suggested, or because the results contradicted received wisdom. "Science today is locked into paradigms," he intoned solemnly. "Every avenue is blocked by beliefs that are wrong, and if you try to get anything published by a journal today, you will run against a paradigm and the editors will turn it down."

Hoyle emphasized that, contrary to certain reports, he did not believe the AIDS virus came from outer space. It "is such a strange virus I have to believe it's a laboratory product," he said. Was Hoyle implying that the pathogen might have been produced by a biological warfare program that went awry? "Yes, that's my feeling," he replied.

Hoyle also suspected that life and indeed the entire universe must be unfolding according to some cosmic plan. The universe is an "obvious fix," Hoyle said. "There are too many things that look accidental which are not." When I asked if Hoyle thought some supernatural intelligence was guiding things, he nodded gravely. "That's the way I look on God. It is a fix, but how it's being fixed I don't know." Of course, many of Hoyle's colleagues—and probably a majority of the public at large—share his view that the universe is, must be, a conspiracy. And perhaps it is. Who knows? But his assertion that scientists would deliberately suppress evidence of space-faring microbes or of legitimate flaws in the big bang theory reveals a fundamental misunderstanding of his colleagues. Most scientists *yearn* for such revolutionary discoveries.

The Sun Principle

Hoyle's eccentricities aside, future observations may prove his skepticism toward the big bang to be at least partially prescient. Astronomers may find that the cosmic microwave background stems not from the flash of the big bang, but from some more mundane source, such as dust in our own Milky Way. The nucleosynthesis evidence, too, may not hold up as well as Schramm and other big bang boosters have claimed. But even if one knocks these two pillars out from under the big bang, the theory can still stand on the red-shift evidence, which even Hoyle agrees provides proof that the universe is expanding.

The big bang theory does for astronomy what Darwin's theory of natural selection did for biology: it provides cohesion, sense, meaning, a unifying narrative. That is not to say that the theory can—or will ever—explain all phenomena. Cosmology, in spite of its close conjunction with particle physics, the most painstakingly precise of sciences, is far from being precise itself. That fact has been demonstrated by the persistent inability of astronomers to agree on a value for the Hubble constant, which is a measure of the size, age, and rate of expansion of the universe. To derive the Hubble constant, one must measure the breadth of the red shift of galaxies and their distance from the earth. The former measurement is straightforward, but the latter is horrendously complicated. Astronomers cannot assume that the apparent brightness of a galaxy is proportional to its distance; the galaxy might be nearby, or it might simply be intrinsically bright. Some astronomers insist that the universe is 10 billion years old or even younger; others are equally sure that it cannot be less than 20 billion years old.[15]

The debate over the Hubble constant offers an obvious lesson: even when performing a seemingly straightforward calculation, cosmologists must make various assumptions that can influence their results; they mus interpret their data, just as evolutionary biologists and historians do. One should thus take with a large grain of salt any claims based on high precision (such as Schramm's assertion that nucleosynthesis calculations agree with theoretical predictions to five decimal points).

More detailed observations of our cosmos will not necessarily resolve questions about the Hubble constant or other issues. Consider: the most mysterious of all stars is our own sun. No one really knows, for example, what causes sunspots or why their numbers wax and wane over periods of roughly a decade. Our ability to describe the universe with simple, elegant models stems in large part from our lack of data, our ignorance. The more clearly we can see the universe in all its glorious detail, the more difficult it will be for us to explain with a simple theory how it came to be that way. Students of human history are well aware of this paradox, but cosmologists may have a hard time accepting it.

This sun principle suggests that many of the more exotic suppositions of cosmology are due for a fall. As recently as the early 1970s black holes were still considered theoretical curiosities, not to be taken seriously. (Einstein himself thought black holes were "a blemish to be removed from his theory by a better mathematical formulation," according to Freeman Dyson.)[16] Gradually, as a result of the proselytizing of John Wheeler and others, they have come to be accepted as real objects. Many theorists are now convinced that almost all galaxies, including our own, harbor gigantic black holes at their cores. The reason for this acceptance is that no one can imagine a better way to explain the violent swirling of matter at the center of galaxies.

These arguments depend on our ignorance. Astronomers should ask themselves this: If they could somehow be whisked to the center of the Andromeda galaxy or our own Milky Way, what would they find there? Would they find something that resembles the black holes described by current theory, or would they encounter something entirely different, something that no one imagined or could have imagined? The sun principle suggests that the latter outcome is more probable. We humans may never see directly into the dust-obscured heart of our own galaxy, let alone into any other galaxy, but we may learn enough to raise doubts about the black hole hypothesis. We may learn enough to learn, once again, how little we know.

The same is true of cosmology in general. We have learned one astounding, basic fact about the universe. We know that the universe is expanding, and may have been for 10 to 20 billion years, just as evolutionary biologists know that all life evolved from a common ancestor through natural selection. But cosmologists are as unlikely to transcend that basic understanding as evolutionary biologists are to leap beyond Darwinism. David Schramm was right. In the future, the late 1980s and early 1990s will be remembered as the golden age of cosmology, when the field achieved a perfect balance between knowledge and ignorance. As more data flood in in years to come, cosmology may become more like botany, a vast collection of empirical facts only loosely bound by theory.

The End of Discovery

Finally, scientists do not enjoy an infinite ability to discover interesting new things about the universe. Martin Harwit, an astrophysicist and historian of science who until 1995 directed the Smithsonian Institution's National Air and Space Museum in Washington, D.C., made this point in his 1981 book, *Cosmic Discovery*:

> The history of most efforts at discovery follows a common pattern, whether we consider the discovery of varieties of insects, the exploration of the oceans for continents and islands, or the search for oil reserves in the ground. There is an initial accelerating rise in the discovery rate as increasing numbers of explorers become attracted. New ideas and tools are brought to bear on the search, and the pace of discovery quickens. Soon, however, the number of discoveries remaining to be made dwindles, and the rate of discovery declines despite the high efficiency of the methods developed. The search is approaching an end. An occasional, previously overlooked feature can be found or a particular rare species encountered; but the rate of discovery begins to decline quickly and then diminishes to a trickle. Interest drops, researchers leave the field, and there is virtually no further activity.[17]

Unlike more experimental fields of science, Harwit pointed out, astronomy is an essentially passive activity. We can only detect celestial phenomena by means of information falling to us from the sky, mostly in the form of electromagnetic radiation. Harwit made various guesses about improvements in mature observational technologies, such as optical telescopes, as well as others that were still nascent, such as gravitational-wave detectors. He presented a graph estimating the rate at which new cosmic discoveries had been made in the past and would be made in the future. The graph was a bell-shaped curve peaking sharply at the year 2000. By that year, we will have discovered roughly half of all the phenomena we can discover, according to Harwit. We will have discovered roughly 90 percent of the accessible phenomena by the year 2200, and the rest will trickle in at a decreasing rate over the next few millennia.

Of course, Harwit conceded, various developments could accelerate or arrest this schedule. "Political factors might dictate that astronomy receive less support in the future. A war might slow the search to a virtual halt, though the postwar era, if there was one, could provide astronomers with discarded military equipment that would again accelerate the discovery rate."[18] Every cloud has a silver lining.

Ironic cosmology will continue, of course, as long as we have poets as imaginative and ambitious as Hawking, Linde, Wheeler, and, yes, Hoyle. Their visions are both humbling, in that they show the limited scope of our empirical knowledge, and exhilarating, since they also testify to the limitlessness of human imagination. At its best, ironic cosmology can keep us awestruck. But it is not science.

John Donne could have been speaking for Hawking, and for all of us, when he wrote: "My thoughts reach all, comprehend all. Inexplicable mystery; I their Creator am in a close prison, in a sick bed, anywhere, and any one of my Creatures, my thoughts is with the Sunne and beyond the Sunne, overtakes the Sunne, and overgoes the Sunne in one pace, one steppe, everywhere."[19] Let that serve as the epitaph of cosmology.

CHAPTER FIVE

The End of
Evolutionary Biology

No other field of science is as burdened by its past as is evolutionary biology. It reeks of what the literary critic Harold Bloom called the anxiety of influence. The discipline of evolutionary biology can be defined to a large degree as the ongoing attempt of Darwin's intellectual descendants to come to terms with his overwhelming influence. Darwin based his theory of natural selection, the central component of his vision, on two observations. First, plants and animals usually produce more offspring than their environment can sustain. (Darwin borrowed this idea from the British economist Thomas Malthus.) Second, these offspring differ slightly from their parents and from each other. Darwin concluded that each organism, in its struggle to survive long enough to reproduce, competes either directly or indirectly with others of its species. Chance plays a role in the survival of any individual organism, but nature will favor, or select, those organisms whose variations make them slightly more fit, that is, more likely to survive long enough to reproduce and pass on those adaptive variations to their offspring.

Darwin could only guess what gives rise to the all-important variations between generations. *On the Origin of Species*, first published in 1859, mentioned a proposal set forth by the French biologist Jean-Baptiste Lamarck, that organisms could pass on not only inherited but also acquired characteristics to their heirs. For example, the constant craning of a giraffe to reach leaves high in a tree would alter its sperm or egg so that its offspring would be born with longer necks. But Darwin was clearly uncomfortable with the idea that adaptation is self-directed. He preferred to think that variations between generations are random, and that only under the pressure of natural selection do they become adaptive and lead to evolution.[1]

Unbeknownst to Darwin, during his lifetime an Austrian monk named Gregor Mendel was conducting experiments that would help refute Lamarck's theory and vindicate Darwin's intuition. Mendel was the first scientist to recognize that natural forms can be subdivided into discrete traits, which are transmitted from one generation to the next by what Mendel termed *hereditary particles* and are now called genes. Genes prevent the blending of traits and thereby preserve them. The recombination of genes that takes place during sexual reproduction, together with occasional genetic mistakes, or mutations, provides the variety needed for natural selection to work its magic.

Mendel's 1868 paper on breeding pea plants went largely unnoticed by the scientific community until the turn of the century. Even then, Mendelian genetics was not immediately reconciled with Darwin's ideas. Some early geneticists felt that genetic mutation and sexual recombination might guide evolution along certain paths independently of natural selection. But in the 1930s and 1940s, Ernst Mayr of Harvard University and other evolutionary biologists fused Darwin's ideas with genetics into a powerful restatement of his theory, called the new synthesis, which affirmed that natural selection is the primary architect of biological form and diversity.

The discovery in 1953 of the structure of DNA—which serves as the blueprint from which all organisms are constructed—confirmed Darwin's intuition that all life is related, descended from a common source. Watson and Crick's finding also revealed the source of both continuity and variation that makes natural selection possible. In addition, molecular biology suggested that all biological phenomena could be explained in mechanical, physical terms.

That conclusion was by no means foregone, according to Gunther Stent. In *The Coming of the Golden Age*, he noted that prior to the unraveling of DNA's structure, some prominent scientists felt that the conventional methods and assumptions of science would prove inadequate for understanding heredity and other basic biological questions. The physicist Niels Bohr was the major proponent of this view. He contended that just as physicists had to cope with an uncertainty principle in trying to understand the behavior of an electron, so would biologists face a fundamental limitation when they tried to probe living organisms too deeply:

> there must remain an uncertainty as regards the physical condition to which [the organism is] subjected, and the idea suggests itself that the minimal freedom we must allow the organism in this respect is just large enough to hide its ultimate secrets from us. On this view, the existence of life must be considered as a starting point in biology, in a similar way as the quantum of action, which appears as an irrational element from the point of view of classical mechanical physics, taken together with the existence of the elementary particles, forms the foundation of quantum mechanics.[2]

Stent accused Bohr of trying to revive the old, discredited concept of vitalism, which holds that life stems from a mysterious essence or force that cannot be reduced to a physical process. But Bohr's vitalist vision has not been borne out. In fact, molecular biology has proved one of Bohr's own dicta, that science, when it is most successful, reduces mysteries to trivialities (although not necessarily comprehensible ones).[3]

What can an ambitious young biologist do to make his or her mark in the post-Darwin, post-DNA era? One alternative is to become more Darwinian than Darwin, to accept Darwinian theory as a supreme insight into nature, one that cannot be transcended. That is the route taken by the arch-clarifier and reductionist Richard Dawkins of the University of Oxford. He has honed Darwinism into a fearsome weapon, one with which he obliterates any ideas that challenge his resolutely materialistic, nonmystical view of life. He seems to view the persistence of creationism and other anti-Darwinian ideas as a personal affront.

I met Dawkins at a gathering convened by his literary agent in Manhattan.[4] He is an icily handsome man, with predatory eyes, a knife-thin nose, and incongruously rosy cheeks. He wore what appeared to be an expensive, custom-made suit. When he held out his finely veined hands to make a point, they quivered slightly. It was the tremor not of a nervous man, but of a finely tuned, high-performance competitor in the war of ideas: Darwin's greyhound.

As in his books, Dawkins in person exuded a supreme self-assurance. His statements often seemed to have an implied preamble: "As any fool can see. . . ." An unapologetic atheist, Dawkins announced that he was not the sort of scientist who thought science and religion addressed separate issues and thus could easily coexist. Most religions, he contended, hold that God is responsible for the design and purpose evident in life. Dawkins was determined to stamp out this point of view. "All purpose comes ultimately from natural selection," he said. "This is the credo that I want to put forward."

Dawkins then spent some 45 minutes setting forth his ultrareductionist version of evolution. He suggested that we think of genes as little bits of software that have only one goal: to make more copies of themselves. Carnations, cheetahs, and all living things are just elaborate vehicles that these "copy-me programs" have created to help them reproduce. Culture, too, is based on copy-me programs, which Dawkins called memes. Dawkins asked us to imagine a book with the message: Believe this book and make your children believe it or when you die you will all go to a very unpleasant place called hell. "That's a very effective piece of copy-me code. Nobody is foolish enough to just accept the injunction, 'Believe this and tell your children to believe it.' You have to be a little more subtle and dress it up in some more elaborate way. And of course we know what I'm talking about." Of course. Christianity, like all religions, is an extremely successful chain letter. What could make more sense?

Dawkins then fielded questions from the audience, a motley assortment of journalists, educators, book editors, and other quasi-intellectuals. One listener was John Perry Barlow, a former Whole Earth hippie and occasional lyricist for the Grateful Dead who had mutated into a New Age cyber-prophet. Barlow, a bearish man with a red bandanna tied around his throat, asked Dawkins a long question having something to do with where information *really* exists.

Dawkins's eyes narrowed, and his nostrils flared ever so slightly as they caught the scent of woolly-headedness. Sorry, he said, but he did not understand the question. Barlow spoke for another minute or so. "I feel you are trying to get at something which interests you but doesn't interest me," Dawkins said and scanned the room for another questioner. Suddenly, the room seemed several degrees chillier.

Later, during a discussion about extraterrestrial life, Dawkins set forth his belief that natural selection is a cosmic principle; wherever life is found, natural selection has been at work. He cautioned that life cannot be too common in the universe, because thus far we have found no evidence of life on other planets in the solar system or elsewhere in the cosmos. Barlow bravely broke in to suggest that our inability to detect alien life-forms may stem from our perceptual inadequacies. "We don't know who discovered water," Barlow added meaningfully, "but we can be pretty sure it wasn't fish." Dawkins turned his level gaze on Barlow. "So you mean we're looking at them all the time," Dawkins asked, "but we don't see them?" Barlow nodded. "Yessss," Dawkins sighed, as if exhaling all hope of enlightening the unutterably stupid world.

Dawkins can be equally harsh with his fellow biologists, those who have dared to challenge the basic paradigm of Darwinism. He has argued, with devastating persuasiveness, that all attempts to modify or transcend Darwin in any significant way are flawed. He opened his 1986 book *The Blind Watchmaker* with the following proclamation: "Our existence once presented the greatest of all mysteries, but . . . it is a mystery no longer because it is solved. Darwin and Wallace solved it, though we shall continue to add footnotes to their solution for a while yet."[5]

"There's always an element of rhetoric in those things," Dawkins replied when I asked him later about the footnotes remark. "On the other hand, it's a legitimate piece of rhetoric," in that Darwin did solve "the mystery of how life came into existence and how life has the beauty, the adaptiveness, the complexity it has." Dawkins agreed with Gunther Stent that all the great advances in biology since Darwin—Mendel's demonstration that genes come in discrete packages, Watson and Crick's discovery of the double-helical structure of DNA—buttressed rather than undermined Darwin's basic idea.

Molecular biology has recently revealed that the process whereby DNA interacts with RNA and proteins is more complicated than previously thought, but the basic paradigm of genetics—DNA-based genetic transmission—is in no danger of collapsing. "What would be a serious reversal," Dawkins said, "would be if you could take a whole organism, a zebra on the Serengeti Plain, and allow it to acquire some characteristic, like learning a new route to the water hole, and have that backward encoded into the genome. Now if anything like that were to happen I really would eat my hat."

There are still some rather large biological mysteries left, such as the origin of life, of sex, and of human consciousness. Developmental biology—which seeks to show how a single fertilized cell becomes a salamander or an evangelist—also raises important issues. "We certainly need to know how that works, and it's going to be very, very complicated." But Dawkins insisted that developmental biology, like molecular genetics before it, would simply fill in more details within the Darwinian paradigm.

Dawkins was "fed up" with those intellectuals who argued that science alone could not answer ultimate questions about existence. "They think science is too arrogant and that there are certain questions that science has no business to ask, that traditionally have been of interest to religious people. As though *they* had any answers. It's one thing to say it's very difficult to know how the universe began, what initiated the big bang, what consciousness is. But if science has difficulty explaining something, there sure as hell is no one else who is going to explain it." Dawkins quoted, with great gusto, a remark by the great British biologist Peter Medawar that some people " 'enjoy wallowing in a nonthreatening squalor of incomprehension.' I want to understand," Dawkins added fiercely, "and understanding means to me scientific understanding."

I asked Dawkins why he thought his message—that Darwin basically told us all we know and all we need to know about life—met with resistance not only from creationists or New Agers or philosophical sophists, but even from obviously competent biologists. "It may be I don't get the point across with sufficient clarity," he replied. But the opposite, of course, is more likely to be true. Dawkins gets his point across with utter clarity, so much so that he leaves no room for mystery, meaning, purpose—or for great scientific revelations beyond the one that Darwin himself gave us.

Gould's Contingency Plan

Naturally, some modern biologists bridle at the notion that they are merely adding footnotes to Darwin's magnum opus. One of Darwin's strongest (in Bloom's sense) descendants is Stephen Jay Gould of Harvard University. Gould has sought to resist the influence of Darwin by denigrating his theory's power, by arguing that it doesn't explain that much. Gould began staking out his philosophical position in the 1960s by attacking the venerable doctrine of uniformitarianism, which holds that the geophysical forces that shaped the earth and life have been more or less constant.[6]

In 1972, Gould and Niles Eldredge of the American Museum of Natural History in New York extended this critique of uniformitarianism to biological evolution by introducing the theory of punctuated equilibrium (also called punk eek or, by critics of Gould and Eldredge, evolution by jerks).[7] New species are only rarely created through the gradual, linear evolution that Darwin described, Gould and Eldredge argued. Rather, speciation is a relatively rapid event that occurs when a group of organisms veers away from its stable parent population and embarks on its own genetic course. Speciation must depend not on the kind of adaptive processes described by Darwin (and Dawkins), but on much more particular, complex, contingent factors.

In his subsequent writings, Gould has hammered away relentlessly at ideas that he claims are implicit in many interpretations of Darwinian theory: progress and inevitability. Evolution does not demonstrate any coherent direction, according to Gould, nor are any of its products—such as *Homo sapiens*—in any sense inevitable; replay the "tape of life" a million times, and this peculiar simian with the oversized brain might never come to be. Gould has also attacked genetic determinism wherever he has found it, whether in pseudoscientific claims about race and intelligence or in much more respectable theories related to sociobiology. Gould packages his skepticism in a prose rich with references to culture high and low and suffused with an acute awareness of its own existence as a cultural artifact. He has been stunningly successful; almost all his books have been bestsellers, and he is one of the most widely quoted scientists in the world.[8]

Before meeting Gould, I was curious about several, seemingly contradictory aspects of his thought. I wondered, for example, just how deep his skepticism, and his aversion to progress, ran. Did he believe, with Thomas Kuhn, that science itself has not demonstrated any coherent progress? Is the course of science as meandering, as aimless, as that of life? Then how did Gould avoid the contradictions that Kuhn fell prey to? Moreover, some critics—and Gould's success ensured that he had legions—accused him of being a crypto-Marxist. But Marx espoused a highly deterministic, progressivist view of history that seemed antithetical to Gould's.

I also wondered whether Gould was backing down on the issue of punctuated equilibrium. In the headline of their original 1972 paper, Gould and Eldredge boldly called punk eek an "alternative" to Darwin's gradualism that might someday supersede it. In the headline of a retrospective essay published in *Nature* in 1993, called "Punctuated Equilibrium Comes of Age," Gould and Eldredge suggested that their hypothesis might be "a useful extension" or "complement" to Darwin's basic model. Punk meek. Gould and Eldredge concluded the 1993 essay with a spasm of disarming honesty. They noted that their theory was only one of many modern scientific ideas emphasizing randomness and discontinuity rather than order and progress. "Punctuated equilibrium, seen in this light, is only paleontology's contribution to a *Zeitgeist*, and *Zeitgeists*, as (literally) transient ghosts of time, should never be trusted. Thus, in developing punctuated equilibrium, we have either been toadies and panderers to fashion, and therefore destined to history's ashheap, or we had a spark of insight about nature's constitution. Only the punctuated and unpredictable future will tell."[9]

I suspected that this uncharacteristic modesty could be traced to events that had transpired in the late 1970s, when journalists were touting punctuated equilibrium as a "revolutionary" advance beyond Darwin. Inevitably, creationists seized on punk eek as proof that the theory of evolution was not universally accepted. Some biologists blamed Gould and Eldredge for having encouraged such claims with their rhetoric. In 1981 Gould tried to set the record straight by testifying at a trial held in Arkansas over whether creationism should be taught alongside evolution in schools. Gould was forced to admit, in effect, that punctuated equilibrium was not a truly revolutionary theory; it was a rather minor technical matter, a squabble among experts.

Gould is disarmingly ordinary-looking. He is short and plump; his face, too, is chubby, adorned with a button nose and graying Charlie Chaplin mustache. When I met him, he was wearing wrinkled khaki pants and an oxford shirt; he looked like the archetypal rumpled, absent-minded professor. But the illusion of ordinariness vanished as soon as Gould opened his mouth. When discussing scientific issues, he talked in a rapid-fire murmur, laying out even the most complex, technical argument with an ease that hinted at much vaster knowledge held in reserve. He decorated his speech, like his writings, with quotations, which he invariably prefaced by saying, "Of course you know the famous remark of. . . ." As he spoke, he often appeared distracted, as if he were not paying attention to his own words. I had the impression that mere speech was not enough to engage him fully; the higher-level programs of his mind roamed ahead, conducting reconnaissance, trying to anticipate possible objections to his discourse, searching for new lines of argument, analogies, quotations. I had the sense that, no matter where I was, Gould was way ahead of me.

Gould acknowledged that his approach to evolutionary biology had been inspired in part by Kuhn's *Structure of Scientific Revolutions*, which he read shortly after it was published in 1962. The book helped Gould believe that he, a young man from "a lower-middle-class family in Queens where nobody had gone to college," might be able to make an important contribution to science. It also led Gould to reject the "inductivist, ameliorative, progressive, add-a-fact-at-a-time-don't-theorize-'til-you're-old model of doing science."

I asked Gould if he believed, as Kuhn did, that science did not advance toward the truth. Shaking his head adamantly, Gould denied that Kuhn held such a position. "I know him, obviously," Gould said. Although Kuhn was the "intellectual father" of the social constructivists and relativists, he nonetheless believed that "there's an objective world out there," Gould asserted; Kuhn felt that this objective world is in *some* sense very hard to define, but he certainly acknowledged that "we have a better sense of what it is now than we did centuries ago."

So was Gould, who had strived so ceaselessly to banish the notion of progress from evolutionary biology, a believer in scientific progress? "Oh sure," he said mildly. "I think all scientists are." No real scientist could possibly be a true cultural relativist, Gould elaborated, because science is too boring. "The day-to-day work of science is *intensely* boring. You've got to clean the mouse cages and titrate your solutions. And you gotta clean your petri dishes." No scientist could endure such tedium unless he or she thought it would lead to "greater empirical adequacy." Gould added, again in reference to Kuhn, that "some people who have large-scale ideas will often express them in an almost strangely exaggerated way just to accentuate the point." (Musing over this remark later, I had to wonder: Was Gould offering a backhanded apology for his own rhetorical excesses?)

Gould glided just as easily past my queries on Marx. He acknowledged that he found some of Marx's proposals quite attractive. For example, Marx's view that ideas are socially embedded and change through conflict, through the clash of theses and antitheses, "is actually a very sensible and interesting theory of change," Gould remarked. "You move by negating the previous one, and then you negate the first negation, and you don't go back to the first one. You've actually moved on somewhere else. I think all of that is quite interesting." Marx's view of social change and revolution, in which "you accumulate small insults to the system until the system itself breaks," is also quite compatible with punctuated equilibrium.

I hardly had to ask the next question: Was Gould, or had he ever been, a Marxist? "You just remember what Marx said," Gould replied before my mouth closed. Marx himself, Gould "reminded" me, once denied that he was a Marxist, because Marxism had become too many things to too many people. No intellectual, Gould explained, wants to identify himself too closely with any "ism," especially one so capacious. Gould also disliked Marx's ideas about progress. "Marx really got caught up in notions of predestiny and determinism, particularly in theories of history, which I think ought to be completely contingent. I *really* think he's dead wrong on that." Darwin, while "too eminent a Victorian to dispense with progress totally," was much more critical of Victorian concepts of progress than Marx had been.

On the other hand, Gould, the adamant antiprogressivist, did not rule out the possibility that culture could display some sort of progress. "Because social inheritance is Lamarckian, there is more of a theoretical basis for belief in progress in culture. It gets derailed all the time by war, *et cetera*, and therefore it becomes contingent. But at least because anything we invent is passed directly to an offspring, there is that possibility of directional accumulation."

When, finally, I asked Gould about punctuated equilibrium, he defended it in spirited fashion. The real significance of the idea, he said, is that "you can't explain [speciation] at the level of the adaptive struggle of the individuals in Darwinian, conventional Darwinian, terms." The trends can only be accounted for by mechanisms operating at the level of species. "You get trends because some species speciate more often, because some species live longer than others," he said. "Since the causes of the birth and death of species are quite different from the causes of the birth and death of organisms, it is a different kind of theory. That's what's interesting. That's where the new theory was in punk eek."

Gould refused to admit that he was in any sense backing down on the issue of punctuated equilibrium or conceding Darwin's supremacy. When I asked him about the switch from "alternative" in his original 1972 paper to "complement" in the 1993 *Nature* retrospective, he exclaimed, "*I* didn't write that!" Gould accused John Maddox, the editor of *Nature*, of inserting "complement" into the paper's headline without checking with him or with Eldredge. "I'm mad at him about that," Gould fumed. But then Gould proceeded to argue that alternative and complement do not really have such different meanings.

"Look, by saying it's an alternative, that doesn't mean that the old kind of gradualism doesn't exist. See, that's another thing I think people miss. The world is full of alternatives, right? I mean we've got men and women, which are alternative states of gender in *Homo sapiens*. I mean if you claim something is an alternative, that doesn't mean it operates exclusively. Gradualism had pretty complete hegemony before we wrote. Here's an alternative to test. I think punctuated equilibrium has an overwhelmingly dominant frequency in the fossil record, which means gradualism exists but it's not really important in the overall pattern of things."

As Gould continued speaking, I began to doubt whether he was really interested in resolving debates over punctuated equilibrium or other issues. When I asked him if he thought biology could ever achieve a final theory, he grimaced. Biologists who hold such beliefs are "naive inductivists," he said. "They actually think that once we sequence the human genome, well, we'll have it!" Even some paleontologists, he admitted, probably think "if we keep at it long enough we really will know the basic features of the history of life and then we'll have it." Gould disagreed. Darwin "had the answer right about the basic interrelationships of organisms, but to me that's only a beginning. It's not over; it's started."

So what did Gould consider the outstanding issues for evolutionary biology? "Oh, there are so many I don't know where to start." He noted that theorists still had to determine the "full panoply of causes" underlying evolution, from molecules on up to large populations of organisms. Then there were "all these contingencies," such as the asteroid impacts that are thought to have caused mass extinctions. "So I would say causes, strengths of causes, levels of causes, and contingency." Gould mused a moment. "That's not a bad formulation," he said, whereupon he took a little notebook from his shirt pocket and scribbled in it.

Then Gould cheerfully rattled off all the reasons that science would *never* answer all these questions. As a historical science, evolutionary biology can offer only retrospective explanations and not predictions, and sometimes it can offer nothing at all because it lacks sufficient data. "If you're missing the evidence of antecedent sequences, then you can't do it at all," he said. "That's why I think we'll never know the origins of language. Because it's not a question of theory; it's a question of contingent history."

Gould also agreed with Gunther Stent that the human brain, created for survival in preindustrial society, is simply not capable of solving certain questions. Research has shown that humans are inept at handling problems that involve probabilities and the interactions of complex variables—such as nature and nurture. "People do not understand that if both genes and culture interact—of *course* they do—you can't then say it's 20 percent genes and 80 percent environment. You can't do that. It's not meaningful. The emergent property is the emergent property and that's all you can ever say about it." Gould was not one of those who invested life or the mind with mystical properties, however. "I'm an old-fashioned materialist," he said. "I think the mind arises from the complexities of neural organization, which we don't really understand very well."

To my surprise, Gould then plunged into a rumination on infinity and eternity. "These are two things that we can't comprehend," he said. "And yet theory almost demands that we deal with it. It's probably because we're not thinking about them right. Infinity is a paradox within Cartesian space, right? When I was eight or nine I used to say, 'Well, there's a brick wall out there.' Well, what's beyond the brick wall? But that's Cartesian space, and even if space is curved you still can't help thinking what's beyond the curve, even if that's not the right way of thinking about it. Maybe all of that's just wrong! Maybe it's a universe of fractal expansions! I don't know what it is. Maybe there are ways in which this universe is structured we just can't think about." Gould doubted whether scientists in any discipline could achieve a final theory, given their tendency to sort things according to preconceived concepts. "I really wonder whether any claim for a final theory isn't just reflecting the way in which we conceptualize them."

Was it possible, given all these limits, that biology, and even science as a whole, might simply go as far as it could and then come to an end? Gould shook his head. "People thought science was ending in 1900, and since then we've got plate tectonics, the genetic basis for life. Why should it stop?" And anyway, Gould added, our theories might reflect our own limitations as truth seekers rather than the true nature of reality. Before I could respond, Gould had already leaped ahead of me. "Of course, if those limits are intrinsic, then science will be complete within the limits. Yeah, yeah. Okay, that's a fair argument. I don't think it's right, but I can understand the structure of it."

Moreover, there may still be great conceptual revolutions in biology's future, Gould argued. "The evolution of life on this planet may turn out to be a very small part of *the* phenomenon of life." Life elsewhere, he elaborated, may very well not conform to Darwinian principles, as Richard Dawkins believed; in fact, the discovery of extraterrestrial life might falsify the claim of Dawkins that Darwin reigns not only here on little earth but throughout the cosmos.

So Gould believes that life exists elsewhere in the universe? I asked. "Don't you?" he retorted. I replied that I thought the issue was entirely a matter of opinion. Gould winced in annoyance; for once, perhaps, he had been caught off guard. Yes, of *course* the existence of life elsewhere is a matter of opinion, he said, but one can still engage in informed speculation. Life seems to have emerged rather readily here on earth, since the oldest rocks that *could* show evidence of life *do* show such evidence. Moreover, "the immensity of the universe and the improbability of absolute uniqueness of any part of it leads to the immense probability that there is some kind of life all over. But we don't know that. Of *course* we don't know that, and I know it's not philosophically incoherent to claim the other."

The key to understanding Gould may be not his alleged Marxism, or liberalism, or anti-authoritarianism, but his fear of his own field's potential for closure. By liberating evolutionary biology from Darwin—and from science as a whole, science defined as the search for universal laws—he has sought to make the quest for knowledge open-ended, even infinite. Gould is far too sophisticated to deny, as some clumsy relativists do, that the fundamental laws uncovered by science exist. Instead, he contends, very persuasively, that the laws do not have much explanatory power; they leave many questions unanswered. He is an extremely adept practitioner of ironic science in its negative capability mode. His view of life can nonetheless be summed up by the old bumper-sticker slogan "Shit happens."

Gould, of course, puts it more elegantly. He noted during our interview that many scientists do not consider history, which deals with particulars and contingency, a part of science. "I think that's a false taxonomy. History is a different type of science." Gould admitted that he found the fuzziness of history, its resistance to straightforward analysis, exhilarating. "I love it! That's because I'm a historian at heart." By transforming evolutionary biology into history—an intrinsically interpretive, ironic discipline, like literary criticism—Gould makes it more amenable for someone with his considerable rhetorical skills. If the history of life is a bottomless quarry of largely random events, he can keep mining it, verbally cherishing one odd fact after another, without ever fearing that his efforts have become trivial or redundant. Whereas most scientists seek to discern the signal underlying nature, Gould keeps drawing attention to the noise. Punctuated equilibrium is not really a theory at all; it is a description of noise.

Gould's great bugaboo is lack of originality. Darwin anticipated the basic concept of punctuated equilibrium in *On the Origin of Species*: "Many species once formed never undergo any further change . . . and the periods during which species have undergone further modification, though long as measured by years, have probably been short in comparison with the periods during which they retain the same form."[10] Ernst Mayr, Gould's colleague at Harvard, proposed in the 1940s that species could appear so rapidly—through the geographic isolation of small populations, for example—that they would leave no trace of the intermediate steps in the fossil record.

Richard Dawkins can find little of value in Gould's oeuvre. To be sure, Dawkins said, speciation may sometimes or even often occur in rapid bursts. But so what? "The important thing is that you have gradualistic selection going on even if that gradualistic selection is telescoped into brief periods around about the time of speciation," Dawkins commented. "So I don't see it as an important point. I see it as an interesting wrinkle on the neo-Darwinian theory."

Dawkins also belittled Gould's insistence that there was no inevitability to the appearance of humans or any other form of intelligent life on earth. "I agree with him on that!" Dawkins said. "So, I think, does everybody else! That's my point! He's tilting at windmills!" Life was single celled for almost three billion years, Dawkins pointed out, and it could well have continued for another three billion without giving rise to multicellular organisms. "So, yes, of *course* there is no inevitability."

Was it possible, I asked Dawkins, that in the long run Gould's view of evolutionary biology would prevail? After all, Dawkins had proposed that the fundamental questions of biology might well be finite, whereas the historical issues Gould addressed were virtually infinite. "If you mean that once all the interesting questions have been solved all you've got to do is find the details," Dawkins replied drily, "I suppose that's got to be true." On the other hand, he added, biologists can never be sure which biological principles have truly universal significance "'til we've been to a few other planets that have life." Dawkins was acknowledging, implicitly, that the deepest questions of biology—To what extent is life on earth inevitable? Is Darwinism a universal or merely terrestrial law?—will never be truly, empirically answered as long as we have only one form of life to study.

The Gaia Heresy

Just as Darwin has absorbed Gould's proposals with barely a burp, so he has metabolized the ideas of another would-be strong scientist, Lynn Margulis of the University of Massachusetts at Amherst. Margulis has challenged what she calls the ultra-Darwinian orthodoxy with several ideas. The first, and most successful, was the concept of symbiosis. Darwin and his heirs had emphasized the role of competition between individuals and species in evolution. In the 1960s, Margulis began arguing that symbiosis had been an equally important factor—and perhaps more important—in the evolution of life. One of the greatest mysteries in evolutionary biology concerns the evolution of prokaryotes, cells that lack a nucleus and are the simplest of all organisms, into eukaryotes, cells that have nuclei. All multicellular organisms, including we humans, consist of eukaryotic cells.

Margulis proposed that eukaryotes may have emerged when one prokaryote absorbed another, smaller one, which became the nucleus. She suggested that such cells should be considered not as individual organisms but as composites. After Margulis was able to provide examples of symbiotic relationships among living microorganisms, she gradually won support for her views on the role of symbiosis in early evolution. She did not stop there, however. Like Gould and Eldredge, she argued that conventional Darwinian mechanisms could not account for the stops and starts observed in the fossil record. Symbiosis, she suggested, could explain why species appear so suddenly and why they persist so long without changing.[11]

Margulis's emphasis on symbiosis led naturally to a much more radical idea: Gaia. The concept and term (Gaia was the Greek goddess of the earth) were originally proposed in 1972 by James Lovelock, a self-employed British chemist and inventor who is perhaps even more iconoclastic than Margulis. Gaia comes in many guises, but the basic idea is that the biota, the sum of all life on earth, is locked in a symbiotic relationship with the environment, which includes the atmosphere, the seas, and other aspects of the earth's surface. The biota chemically regulates the environment in such a way as to promote its own survival. Margulis was immediately taken with the idea, and she and Lovelock have collaborated in promulgating it ever since.[12]

I met Margulis in May 1994 in the first-class lounge of New York's Pennsylvania Station, where she was waiting for a train. She resembled an aging tomboy: she had short hair and ruddy skin, and she wore a striped, short-sleeved shirt and khaki pants. She dutifully played the radical, at first. She ridiculed the suggestion of Dawkins and other ultra-Darwinians that evolutionary biology might be nearing completion. "*They're* finished," Margulis declared, "but that's just a small blip in the twentieth century history of biology rather than a full-fledged and valid science."

She emphasized that she had no problem with the basic premise of Darwinism. "Evolution no doubt occurs, and it's been seen to occur, and it's occurring now. Everyone who's scientific minded agrees with that. The question is, how does it occur? And that's where everyone parts company." Ultra-Darwinians, by focusing on the gene as the unit of selection, had failed to explain how speciation occurs; only a much broader theory that incorporated symbiosis and higher-level selection could account for the diversity of the fossil record and of life today, according to Margulis.

Symbiosis, she added, also allows a kind of Lamarckianism, or inheritance of acquired characteristics. Through symbiosis, one organism can genetically absorb or infiltrate another and thereby become more fit. For example, if a translucent fungus absorbs an alga that can perform photosynthesis, the fungus may acquire the capacity for photosynthesis, too, and pass it on to its offspring. Margulis said that Lamarck had been unfairly cast as the goat of evolutionary biology. "We have this British–French business. Darwin's all right and Lamarck is bad. It's really terrible." Margulis noted that symbiogenesis, the creation of new species through symbiosis, was not really an original idea. The concept was first proposed late in the last century, and it has been resurrected many times since then.

Before meeting Margulis, I had read a draft of a book she was writing with her son, Dorion Sagan, called *What Is Life?* The book was an amalgam of philosophy, science, and lyric tributes to "life: the eternal enigma." It argued, in effect, for a new holistic approach to biology, in which the animist beliefs of the ancients would be fused with the mechanistic views of post-Newton, post-Darwin science.[13] Margulis conceded that the book was aimed less at advancing testable, scientific assertions than at encouraging a new philosophical outlook among biologists. But the only difference between her and biologists like Dawkins, she insisted, was that she admitted her philosophical outlook instead of pretending that she didn't have one. "Scientists are no cleaner with respect to being untouched by culture than anyone else."

Did that mean she did not believe that science could achieve absolute truth? Margulis pondered the question a moment. She noted that science derives its power and persuasiveness from the fact that its assertions can be checked against the real world—unlike the assertions of religion, art, and other modes of knowledge. "But I don't think that's the same as saying there's absolute truth. I don't think there's absolute truth, and if there is, I don't think any person has it."

But then, perhaps realizing how close she was edging toward relativism—toward being what Harold Bloom had called a mere rebel—Margulis took pains to steer herself back toward the scientific mainstream. She said that, although she was often considered a feminist, she was not and resented being typecast as one. She conceded that, in comparison to such concepts as "survival of the fittest" and "nature red in tooth and claw," Gaia and symbiosis might *seem* feminine. "There is that cultural overtone, but I consider that just a complete distortion."

She rejected the notion—often associated with Gaia—that the earth is in some sense a living organism. "The earth is *obviously* not a live organism," Margulis said, "because no single living organism cycles its waste. That's *so* anthropomorphic, *so* misleading." James Lovelock encouraged this metaphor, she claimed, because he thought it would aid the cause of environmentalism, and because it suited his own quasi-spiritual leanings. "He says it's an okay metaphor because it's better than the old one. I think it's bad because it's just getting the scientists mad at you, because you're encouraging irrationality." (Actually, Lovelock, too, has reportedly voiced doubts about some of his earlier claims concerning Gaia and has even thought of renouncing the term.)[14]

Both Gould and Dawkins have ridiculed Gaia as pseudoscience, poetry posing as a theory. But Margulis is, in at least one sense, much more hard-nosed, more of a positivist, than they are. Gould and Dawkins resorted to speculation about extraterrestrial life in order to buttress their views of life on earth. Margulis scoffed at these tactics. Any proposals concerning the existence of life elsewhere in the universe—or its Darwinian or non-Darwinian nature—are sheer speculation, she said. "You have no constraints on the answer to that, whether it's a frequent or infrequent thing. So I don't see how people can have strong opinions on that. Let me put it this way: opinions aren't science. There's no scientific basis! It's just opinion!"

She remembered that in the early 1970s she had received a call from the director Steven Spielberg, who was in the process of writing the movie *ET*. Spielberg asked Margulis if she thought it was likely or even possible that an extraterrestrial would have two hands, each with five fingers. "I said, 'You're making a movie! Just make it fun! What the hell do you care! Don't try to confuse yourself that it's science!' "

Toward the end of our interview, I asked Margulis if she minded always being referred to as a provocateur or gadfly, or as someone who was "fruitfully wrong," as one scientist had put it.[15] She pressed her lips together, brooding over the question. "It's kind of dismissive, not serious," she replied. "I mean, you wouldn't do this to a serious scientist, would you?" She stared at me, and I finally realized her question was not rhetorical; she really wanted an answer. I agreed that the descriptions seemed somewhat condescending.

"Yeah, that's right," she mused. Such criticism did not bother her, she insisted. "Anyone who makes this kind of *ad hominem* criticism exposes himself, doesn't he? I mean, if their argument is just based on provocative adjectives about me rather than the substance of the issue, then . . ." Her voice trailed off. Like so many strong scientists, Margulis cannot help but yearn, now and then, to be simply a respected member of the status quo.

Kauffman's Passion for Order

Perhaps the most ambitious and radical modern challenger of Darwin is Stuart Kauffman, a biochemist at the Santa Fe Institute, headquarters of the ultratrendy field of complexity (see Chapter 8). In the 1960s, when he was still a graduate student, Kauffman began to suspect that Darwin's theory of evolution was seriously flawed, in that it could not account for the seemingly miraculous ability of life to appear and then to perpetuate itself in such marvelous ways. After all, the second law of thermodynamics decrees that everything in the universe is drifting inexorably toward "heat death," or universal blandness.

Kauffman tested his ideas by simulating the interaction of various abstract agents—supposedly representative of chemical and biological substances—on computers. He came to several conclusions. One was that when a system of simple chemicals reaches a certain level of complexity, it undergoes a dramatic transition, akin to the phase change that occurs when liquid water freezes. The molecules begin spontaneously combining to create larger molecules of increasing complexity and catalytic capability. Kauffman argued that this process of self-organization or autocatalysis, rather than the fortuitous formation of a molecule with the ability to replicate and evolve, led to life.

Kauffman's other hypothesis was even farther reaching, in that it challenged perhaps the central tenet of biology: natural selection. According to Kauffman, complex arrays of interacting genes subject to random mutations do not evolve randomly. Instead, they tend to converge toward a relatively small number of patterns, or attractors, to use the term favored by chaos theorists. In his 709-page 1993 book, *The Origins of Order: Self-Organization and Selection in Evolution*, Kauffman contended that this ordering principle, which he sometimes called antichaos, may have played a larger role than natural selection in guiding the evolution of life, particularly as life grew in complexity.[16]

Kauffman, when I first met him during a visit to Santa Fe in May 1994, had a broad, deeply tanned face and wavy gray hair thinning at the crown. He wore standard-issue Santa Fe apparel: denim shirt, khakis, hiking boots. He seemed at once tremulous and vulnerable and supremely confident. His speech, like the improvisations of a jazz musician, was short on melody, long on digression. Like a salesperson trying to establish bonhomie, he kept calling me John. He clearly enjoyed talking about philosophy. Over the course of our conversation he presented minilectures not only on his antichaos theories but also on the limits of reductionism, the difficulty of falsifying theories, and the social context of scientific facts.

Early in our conversation, Kauffman recalled an article I had written for *Scientific American* on the origin of life.[17] In the article I had quoted Kauffman saying, of his own origin-of-life theory, "I'm sure I'm right." Kauffman told me he had been embarrassed to see this quotation in print; he had vowed to avoid such hubris in the future. Somewhat to my regret, Kauffman kept his word, for the most part. During our conversation, he took great pains to qualify his assertions: "Now I'm not going to say that I'm sure I'm right, John, but . . ."

Kauffman had just finished a book, *At Home in the Universe*, that spelled out the implications of his theories on biological evolution. "The line that I take in my Oxford book, John, that I think is right—I mean, I think it's right in the sense that it's all perfectly plausible; *much* to be shown experimentally. But at least in terms of mathematical models, here you have a body of models that says the emergence of life might be a natural phenomenon, in the sense that, given a sufficiently complicated set of reacting molecules, you'd expect to crystallize autocatalytic subsets. So if that view is right, as I told you with abundant enthusiasm a couple of years ago"—big smile—"then we're not incredibly improbable accidents." In fact, life almost certainly exists elsewhere in the universe, he added. "And therefore we are at home in the universe in a different way than we would be if life were this incredibly improbable event that happened on one planet and one planet only because it was so improbable that you wouldn't expect it to happen at all."

Kauffman put the same spin on his theory about how networks of genes tend to settle into certain recurring patterns. "Again, suppose that I'm right," Kauffman said; then much of the order displayed by biological systems results not from "the hard-won success of natural selection" but from these pervasive order-generating effects. "The whole point of it is that it's spontaneous order. It's order for free, okay? Once again, if that view is right, then not only do we have to modify Darwinism to account for it, but we understand something about the emergence and order of life in a different way."

His computer simulations offer another, more sobering message,

Kauffman said. Just as the addition of a single grain of sand to a large sandpile can trigger avalanches down its sides, so can a change in the fitness of one species cause a sudden change in the fitness of all the other species in the ecosystem, which can culminate in an avalanche of extinctions. "To say it metaphorically, the best adaptation each of us achieves may unleash an avalanche that leads to our ultimate demise, okay? Because we're all playing the game together and sending out ripples into the system we mutually create. Now *that* bespeaks humility." Kauffman credited the sandpile analogy to Per Bak, a physicist associated with the Santa Fe Institute who had developed a theory called self-organized criticality (which is discussed in Chapter 8).

When I brought up my concern that many scientists, and those at the Santa Fe Institute in particular, seemed to confuse computer simulations with reality, Kauffman nodded. "I agree with you. That personally bothers me a lot," he replied. Viewing some simulations, he said, "I cannot tell where the boundary is between talking about *the world*—I mean, everything out there—and really neat computer games and art forms and toys." When *he* did computer simulations, however, he was always "trying to figure out how something in the world works, or almost always. Sometimes I'm busy just trying to find things that just seem interesting, and I wonder if they apply. But I don't think you're doing science unless what you're doing winds up fitting something out there in the world, demonstrably. And that ultimately means being testable."

His model of genetic networks "makes all *kinds* of predictions" that will probably be tested within the next 15 or 20 years, Kauffman said. "It's testable with some caveats. When you've got a system with 100,000 components, and you can't take the system apart in detail—yet—what are the appropriate testable consequences? They're going to have to be statistical consequences, right?"

How could his origin-of-life theory be tested? "You could be asking two different questions," Kauffman replied. In one sense, the question concerns the way in which life actually arose on earth some four billion years ago. Kauffman did not know whether his theory, or any theory, could satisfactorily address this historical question. On the other hand, one might test his theory by trying to create autocatalytic sets in the laboratory. "Tell you what. We'll make a deal. Whether I do it or somebody else does it, if somebody makes collectively autocatalytic sets of molecules with the phase transitions and reaction graphs, you owe me a dinner, okay?"

There are parallels between Kauffman and other challengers of the status quo in evolutionary biology. First, Kauffman's ideas have historical precedents, just as punctuated equilibrium and symbiosis do. Kant, Goethe, and other pre-Darwinian thinkers speculated that general mathematical principles or rules might underlie the patterns of nature. Even after Darwin, many biologists remained convinced of the existence of some order-generating force in addition to natural selection that counteracts the universal drift toward thermodynamic sameness and gives rise to biological order. Twentieth-century proponents of this viewpoint, which is sometimes called rational morphology, include D'Arcy Wentworth Thompson, William Bateson, and, most recently, Brian Goodwin.[18]

Moreover, Kauffman seems to be motivated at least as much by philosophical conviction about how things must be as by scientific curiosity about how things really are. Gould stresses the importance of chance, contingency, in shaping evolution. Margulis eschews the reductionism of neo-Darwinism for a more holistic approach. Kauffman, similarly, feels that accident alone cannot have created life; our cosmos must harbor some fundamental order-generating tendency.

Finally, Kauffman, like Gould and Margulis, has struggled to define his relationship to Darwin. In his interview with me he said he viewed antichaos as a complement to Darwinian natural selection. At other times, he has proclaimed that antichaos is the primary factor in evolution and that natural selection's role has been minor or nonexistent. Kauffman's continued ambivalence over this issue was starkly revealed in a typeset draft of *At Home in the Universe*, which he gave me in the spring of 1995. On the book's first page, Kauffman proclaimed that Darwinism was "wrong," but he had crossed out "wrong" and replaced it with "incomplete." Kauffman went back to "wrong" in the galleys of his book, released several months later. What did the final, published version say? "Incomplete."

Kauffman has one powerful ally in Gould, who proclaimed on the cover of Kauffman's *Origins of Order* that it would become "a landmark and a classic as we grope towards a more comprehensive and satisfying theory of evolution." This is a strange alliance. Whereas Kauffman has argued that the laws of complexity he discerns in his computer simulations have lent the evolution of life a certain inevitability, Gould has devoted his career to arguing that virtually *nothing* in the history of life was inevitable. In his conversation with me, Gould also specifically repudiated the proposal that the history of life unfolded according to mathematical laws. "It's a very deep position," Gould said, "but I also think it's very deeply wrong." What Gould and Kauffman do have in common is that each has challenged the assertion of Richard Dawkins and other hard-core Darwinians that evolutionary theory has already more or less explained the history of life. In providing blurbs for Kauffman's books, Gould shows that he adheres to the old maxim "The enemy of my enemy is my friend."

For the most part, Kauffman has had little success at winning a following for his ideas. Perhaps the major problem is that his theories are statistical in nature, as he himself admits. But one cannot confirm a statistical prediction about the probability of life's origin and its subsequent evolution when one only has one data point—terrestrial life—to consider. One of the harshest assessments of Kauffman's work comes from John Maynard Smith, a British evolutionary biologist who, like Dawkins, is renowned for his sharp tongue and who pioneered the use of mathematics in evolutionary biology. Kauffman once studied under Maynard Smith, and he has spent countless hours trying to convince his former mentor of the importance of his work—apparently in vain. In a public debate in 1995, Maynard Smith said of self-organized criticality, the sandpile model advanced by Per Bak and embraced by Kauffman, "I just find the whole enterprise contemptible." Maynard Smith told Kauffman later, over beers, that he did not find Kauffman's approach to biology interesting.[19] To a practitioner of ironic science such as Kauffman, there can be no crueler insult.

Kauffman is at his most eloquent and persuasive when he is in his critical mode. He has implied that the evolutionary theory promulgated by biologists such as Dawkins is cold and mechanical; it does not do justice to the majesty and mystery of life. Kauffman is right; there *is* something unsatisfying, tautological, about Darwinian theory, even when it is explicated by a rhetorician as skilled as Dawkins. But Dawkins, at least, distinguishes between living and nonliving things. Kauffman seems to see all phenomena, from bacteria to galaxies, as manifestations of abstract mathematical forms that undergo endless permutations. He is a mathematical aesthete. His vision is similar to that of the particle physicists who call God a geometer. Kauffman has suggested that his vision of life is more meaningful and comforting than that of Dawkins. But most of us, I suspect, can identify more with Dawkins's pushy little replicators than we can with Kauffman's Boolean functions in N-space. Where is the meaning and comfort in these abstractions?

The Conservatism of Science

One constant threat to the status quo in science—and in all human endeavors—is the desire of new generations to make their mark on the world. Society also has an insatiable appetite for the new. These twin phenomena are largely responsible for the rapid turnover of styles in the arts, where change for its own sake is embraced. Science is hardly immune to these influences. Gould's punctuated equilibrium, Margulis's Gaia, and Kauffman's antichaos have all had their minute of fame. But it is much more difficult to achieve lasting change in science than in the arts, for obvious reasons. Science's success stems in large part from its conservatism, its insistence on high standards of effectiveness. Quantum mechanics and general relativity were as new, as surprising, as anyone could ask for. But they were believed ultimately not because they imparted an intellectual thrill but because they were effective: they accurately predicted the outcome of experiments. Old theories are old for good reason. They are robust, flexible. They have an uncanny correspondence to reality. They may even be True.

Would-be revolutionaries face another problem. The scientific culture was once much smaller and therefore more susceptible to rapid change. Now it has become a vast intellectual, social, and political bureaucracy, with inertia to match. Stuart Kauffman, during one of our conversations, compared the conservatism of science to that of biological evolution, in which history severely constrains change. Not only science, but many other systems of ideas—and particularly those with important social consequences—tend to "stabilize and freeze in" over time, Kauffman noted. "Think of the evolution of standard operating procedure on ships, or aircraft carriers," he said. "It's an incredibly conservative process. If you wandered in and tried to design, *ab initio* [from scratch], procedures on an aircraft carrier, I mean, you'd just blow it to shit!"

Kauffman leaned toward me. "This is really interesting," he said. "Take the law, okay? British common law has evolved for what, 1,200 years? There is this enormous corpus with a whole bunch of concepts about what constitutes reasonable behavior. It would be really hard to change all that! I wonder if you could show that as a web of concepts matures in any area— in all these cases we're making maps to make our way in the world—I wonder if you could show that somehow the center of it got more and more resistant to change." Oddly enough, Kauffman was presenting an excellent argument why his own radical theories about the origin of life and of biological order would probably never be accepted. If any scientific idea has proved its ability to overcome all challengers, it is Darwin's theory of evolution.

The Mysterious Origin of Life

If I were a creationist, I would cease attacking the theory of evolution— which is so well supported by the fossil record—and focus instead on the origin of life. This is by far the weakest strut of the chassis of modern biology. The origin of life is a science writer's dream. It abounds with exotic scientists and exotic theories, which are never entirely abandoned or accepted, but merely go in and out of fashion.[20]

One of the most diligent and respected origin-of-life researchers is Stanley Miller. He was a 23-year-old graduate student in 1953 when he sought to recreate the origin of life in a laboratory. He filled a sealed glass apparatus with a few liters of methane, ammonia, and hydrogen (representing the atmosphere) and some water (the ocean). A spark-discharge device zapped the gases with simulated lightning, while a heating coil kept the waters bubbling. Within a few days, the water and gases were stained with a reddish goo. On analyzing the substance, Miller found to his delight that it was rich in amino acids. These organic compounds are the building blocks of proteins, the basic stuff of life.

Miller's results seemed to provide stunning evidence that life could arise from what the British chemist J. B. S. Haldane had called the "primordial soup." Pundits speculated that scientists, like Mary Shelley's Dr. Franken-stein, would shortly conjure up living organisms in their laboratories and thereby demonstrate in detail how genesis unfolded. It hasn't worked out that way. In fact, almost 40 years after his original experiment, Miller told me that solving the riddle of the origin of life had turned out to be more difficult than he or anyone else had envisioned. He recalled one prediction, made shortly after his experiment, that within 25 years scientists would "surely" know how life began. "Well, 25 years have come and gone," Miller said drily.

After his 1953 experiment, Miller dedicated himself to the search for the secret of life. He developed a reputation as both a rigorous experimentalist and a bit of a curmudgeon, someone who is quick to criticize what he feels is shoddy work. When I met Miller in his office at the University of California at San Diego, where he is a professor of biochemistry, he fretted that his field still had a reputation as a fringe discipline, not worthy of serious pursuit. "Some work is better than others. The stuff that is awful does tend to drag it down. I tend to get very upset about that. People do good work, and then you see this garbage attract attention." Miller seemed unimpressed with any of the current proposals on the origin of life, referring to them as "nonsense" or "paper chemistry." He was so contemptuous of some hypotheses that, when I asked his opinion of them, he merely shook his head, sighed deeply, and snickered—as if overcome by the folly of humanity. Stuart Kauffman's theory of autocatalysis fell into this category. "Running equations through a computer does not constitute an experiment," Miller sniffed.

Miller acknowledged that scientists may never know precisely where and when life emerged. "We're trying to discuss a historical event, which is very different from the usual kind of science, and so criteria and methods are very different," he remarked. But when I suggested that Miller sounded pessimistic about the prospects for discovering life's secret, he looked appalled. Pessimistic? Certainly not! He was optimistic!

One day, he vowed, scientists would discover the self-replicating molecule that had triggered the great saga of evolution. Just as the discovery of the microwave afterglow of the big bang legitimized cosmology, so would the discovery of the first genetic material legitimize Miller's field. "It would take off like a rocket," Miller muttered through clenched teeth. Would such a discovery be immediately self-apparent? Miller nodded. "It will be in the nature of something that will make you say, 'Jesus, there it is. How could you have overlooked this for so long?' And everybody will be totally convinced."

When Miller performed his landmark experiment in 1953, most scientists still shared Darwin's belief that proteins were the likeliest candidates for self-reproducing molecules, since proteins were thought to be capable of reproducing and organizing themselves. After the discovery that DNA is the basis for genetic transmission and for protein synthesis, many researchers began to favor nucleic acids over proteins as the ur-molecules. But there was a major hitch in this scenario. DNA can make neither proteins nor copies of itself without the help of catalytic proteins called enzymes. This fact turned the origin of life into a classic chicken-or-egg problem: which came first, proteins or DNA?

In *The Coming of the Golden Age*, Gunther Stent, prescient as always, suggested that this conundrum could be solved if researchers found a self-replicating molecule that could act as its own catalyst.[21] In the early 1980s, researchers identified just such a molecule: ribonucleic acid, or RNA, a single-strand molecule that serves as DNA's helpmate in manufacturing proteins. Experiments revealed that certain types of RNA could act as their own enzymes, snipping themselves in two and splicing themselves back together again. If RNA could act as an enzyme then it might also be able to replicate itself without help from proteins. RNA could serve as both gene and catalyst, egg and chicken.

But the so-called RNA-world hypothesis suffers from several problems. RNA and its components are difficult to synthesize under the best of circumstances, in a laboratory, let alone under plausible prebiotic conditions. Once RNA is synthesized, it can make new copies of itself only with a great deal of chemical coaxing from the scientist. The origin of life "has to happen under easy conditions, not ones that are very special," Miller said. He is convinced that some simpler—and possibly quite dissimilar—molecule must have paved the way for RNA.

Lynn Margulis, for one, doubts whether investigations of the origin of life will yield the kind of simple, self-validating answer of which Miller dreams. "I think that may be true of the cause of cancer but not of the origin of life," Margulis said. Life, she pointed out, emerged under complex environmental conditions. "You have day and night, winter and summer, changes in temperature, changes in dryness. These things are historical accumulations. Biochemical systems are effectively historical accumulations. So I don't think there is ever going to be a packaged recipe for life: add water and mix and get life. It's not a single-step process. It's a cumulative process that involves a lot of changes." The smallest bacterium, she noted, "is so much more like people than Stanley Miller's mixtures of chemicals, because it already has these system properties. So to go from a bacterium to people is less of a step than to go from a mixture of amino acids to that bacterium."

Francis Crick wrote in his book *Life Itself* that "the origin of life appears to be almost a miracle, so many are the conditions which would have to be satisfied to get it going."[22] (Crick, it should be noted, is an agnostic leaning toward atheism.) Crick proposed that aliens visiting the earth in a spacecraft billions of years ago may have deliberately seeded it with microbes.

Perhaps Stanley Miller's hope will be fulfilled: scientists will find some clever chemical or combination of chemicals that can reproduce, mutate, and evolve under plausible prebiotic conditions. The discovery would be sure to launch a new era of applied chemistry. (The vast majority of researchers focus on this goal, rather than on the elucidation of life's origin.) But given our lack of knowledge about the conditions under which life began, any theory of life's origin based on such a finding would always be subject to doubts. Miller has faith that biologists will know the answer to the riddle of life's origin when they see it. But his belief rests on the premise that the answer will be plausible, if only retrospectively. Who said the origin of life on earth was plausible? Life might have emerged from a freakish convergence of improbable and even unimaginable events.

Moreover, the discovery of a plausible ur-molecule, when or if it happens, is unlikely to tell us what we really want to know: Was life on earth inevitable or a freak occurrence? Has it happened elsewhere or only in this lonely, lonely spot? These questions can only be resolved if we discover life beyond the earth. Society seems increasingly reluctant to underwrite such investigations. In 1993, Congress shut down NASA's SETI (Search for Extraterrestrial Intelligence) program, which scanned the heavens for radio signals generated by other civilizations. The dream of a manned mission to Mars, the most likely site for extraterrestrial life in the solar system, has been indefinitely deferred.

Even so, scientists may find evidence of life beyond the earth tomorrow. Such a discovery would transform all of science and philosophy and human thought. Stephen Jay Gould and Richard Dawkins might be able to settle their argument over whether natural selection is a cosmic or merely terrestrial phenomenon (although each would doubtless find ample evidence for his point of view). Stuart Kauffman might be able to determine whether the laws he discerns in his computer simulations prevail in the real world. If the extraterrestrials are intelligent enough to have developed their own science, Edward Witten may learn whether superstring theory really is the inevitable culmination of any search for the fundamental rules governing reality. Science fiction will become fact. The *New York Times* will resemble the *Weekly World News*, one of the supermarket tabloids that prints "photographs" of presidents hobnobbing with aliens. One can always hope.

CHAPTER SIX

The End of Social Science

Everything would have been easy for Edward O. Wilson if he had just stuck to ants. Ants lured him into biology when he was a boy growing up in Alabama, and they remain his greatest source of inspiration. He has written stacks of papers and several books on the tiny creatures. Ant colonies line Wilson's office at Harvard University's Museum of Comparative Zoology. Showing them off to me, he was as proud and excited as a 10-year-old child. When I asked Wilson if he had exhausted the topic of ants yet, he exulted, "We're only just beginning!" He had recently embarked on a survey of *Pheidole*, one of the most abundant genera in the animal kingdom. *Pheidole* is thought to include more than 2,000 species of ants, most of which have never been described or even named. "I guess with that same urge that makes men in their middle age decide that at last they are going to row across the Atlantic in a rowboat or join a group to climb K2, I decided that I would take on *Pheidole*," Wilson said.[1]

Wilson was a leader in the effort to conserve the earth's biological diversity, and his grand goal was to make *Pheidole* a benchmark of sorts for biologists seeking to monitor the biodiversity of different regions. Drawing on Harvard's collection of ants, the largest in the world, Wilson was generating a set of painstaking pencil drawings of each species of *Pheidole* along with descriptions of its behavior and ecology. "It probably looks crushingly dull to you," Wilson apologized as he flipped through his drawings of *Pheidole* species (which were actually compellingly monstrous). "To me it's one of the most satisfying activities imaginable." He confessed that, when he peered through his microscope at a previously unknown species, he had "the sensation of maybe looking upon—I don't want to get too poetic—of looking upon the face of creation." A single ant was enough to render Wilson awestruck before the universe.

I first detected a martial spirit glinting through Wilson's boyish exuberance when we walked over to the ant farm sprawling across a counter in his office. These are leaf-cutter ants, Wilson explained, which range from South America as far north as Louisiana. The scrawny little specimens scurrying across the surface of the spongelike nest are the workers; the soldiers lurk within. Wilson pulled a plug from the top of the nest and blew into the hole. An instant later several bulked-up behemoths boiled to the surface, BB-sized heads tossing, mandibles agape. "They can cut through shoe leather," Wilson remarked, a bit too admiringly. "If you tried to dig into a leaf-cutter nest, they would gradually dispatch you, like a Chinese torture, by a thousand cuts." He chuckled.

Wilson's pugnacity—innate or inculcated?—emerged more clearly later on, when he discussed the continued reluctance of American society to confront the role played by genes in shaping human behavior. "This country is so seized by our civic religion, egalitarianism, that it just averts its gaze from anything that would seem to detract from that central ethic we have that everybody is equal, that perfect societies can be built with the good will of people." As he delivered this sermon, Wilson's long-boned, Yankee-farmer face, usually so genial, became as stony as a Puritan preacher's.

There are two—at least two—Edward Wilsons. One is the poet of social insects and the passionate defender of all the earth's biodiversity. The other is a fiercely ambitious, competitive man struggling with his sense that he is a latecomer, that his field is more or less complete. Wilson's response to the anxiety of influence was quite different from that of Gould, Margulis, Kauffman, and others who have wrestled with Darwin. For all their differences, they responded to Darwin's dominance by arguing that Darwinian theory has limited explanatory power, that evolution is much more complicated than Darwin and his modern heirs have suggested. Wilson took the opposite course. He sought to extend Darwinism, to show that it could explain more than anyone—even Richard Dawkins—had thought.

Wilson's role as the prophet of sociobiology can be traced back to a crisis of faith he suffered in the late 1950s, just after his arrival at Harvard. Although he was already one of the world's authorities on social insects, he began to brood over the apparent insignificance—at least in the eyes of other scientists—of his field of research. Molecular biologists, exhilarated by their discovery of the structure of DNA, the basis of genetic transmission, had begun questioning the value of studying whole organisms, such as ants. Wilson once recalled that James Watson, who was then at Harvard and still flushed with the excitement of having discovered the double helix, "radiated contempt" for evolutionary biology, which he saw as a glorified version of stamp collecting rendered obsolete by molecular biology.[2]

Wilson responded to this challenge by broadening his outlook, by seeking the rules of behavior governing not only ants but all social animals. That effort culminated in *Sociobiology: The New Synthesis*. Published in 1975, the book was for the most part a magisterial survey of nonhuman social animals, from ants and termites to antelope and baboons. Drawing on ethology, population genetics, and other disciplines, Wilson showed how mating behavior and division of labor could be understood as adaptive responses to evolutionary pressure.

Only in the last chapter did Wilson turn to humans. He drew attention to the obvious fact that sociology, the study of human social behavior, was badly in need of a unifying theory. "There have been attempts at system building but . . . they came to little. Much of what passes for theory in sociology today is really labeling of phenomena and concepts, in the expected manner of natural history. Process is difficult to analyze because the fundamental units are elusive, perhaps nonexistent. Syntheses commonly consist of the tedious cross-referencing of differing sets of definitions and metaphors erected by the more imaginative thinkers."[3]

Sociology would only become a truly scientific discipline, Wilson argued, if it submitted to the Darwinian paradigm. He noted, for example, that warfare, xenophobia, the dominance of males, and even our occasional spurts of altruism could all be understood as adaptive behaviors stemming from our primordial compulsion to propagate our genes. In the future, Wilson predicted, further advances in evolutionary theory as well as in genetics and neuroscience would enable sociobiology to account for a wide variety of human behavior; sociobiology would eventually subsume not only sociology, but also psychology, anthropology, and all the "soft" social sciences.

The book received generally favorable reviews. Yet some scientists, including Wilson's colleague Stephen Jay Gould, rebuked Wilson for suggesting that the human condition was somehow *inevitable*. Wilson's views, critics argued, represented an updated version of social Darwinism; that notorious Victorian-era doctrine, by conflating what is with what should be, had provided a scientific justification for racism, sexism, and imperialism. The attacks on Wilson peaked in 1978 at a meeting of the American Association for the Advancement of Science. A member of a radical group called the International Committee Against Racism dumped a pitcher of water on Wilson's head while shouting, "You're all wet."[4]

Undeterred, Wilson went on to write two books on human sociobiology with Charles Lumsden, a physicist at the University of Toronto: *Genes, Mind, and Culture* (1981) and *Promethean Fire* (1983). Wilson and Lumsden conceded in the latter book "the sheer difficulty of creating an accurate portrayal of genetic and cultural interaction." But they declared that the way to cope with this difficulty was not to continue "the honored tradition of social theory written as literary criticism," but to create a rigorous mathematical theory of the interaction between genes and culture. "The theory we wished to build," Wilson and Lumsden wrote, "would contain a system of linked abstract processes expressed as far as possible in the form of explicit mathematical structures that translate the processes back to the real world of sensory experience."[5]

The books that Wilson wrote with Lumsden were not as well received as *Sociobiology* had been. One critic recently called their view of human nature "grimly mechanistic" and "simplistic."[6] During our interview, Wilson was nonetheless more bullish than ever about the prospects for sociobiology. While granting that support for his proposals was very slim in the 1970s, he insisted that "a lot more evidence exists today" that many human traits—ranging from homosexuality to shyness—have a genetic basis; advances in medical genetics have also made genetic explanations of human behavior more acceptable to scientists and to the public. Human sociobiology has flourished not only in Europe, which has a Sociobiological Society, but also in the United States, Wilson said; although many scientists in the United States shun the term *sociobiology* because of its political connotations, disciplines with names such as biocultural studies, evolutionary psychology, and Darwinian studies of human behavior are all "sprigs" growing from the trunk of sociobiology.[7]

Wilson was still convinced that sociobiology would eventually subsume not only the social sciences but also philosophy. He was writing a book, tentatively titled *Natural Philosophy*, about how findings from sociobiology would help to resolve political and moral issues. He intended to argue that religious tenets can and should be "empirically tested" and rejected if they are incompatible with scientific truths. He suggested, for example, that the Catholic church might examine whether its prohibition against abortion—a dogma that contributes to overpopulation—conflicts with the larger moral goal of preserving all the earth's biodiversity. As Wilson spoke, I recalled one colleague's comment that Wilson combined great intelligence and learnedness with, paradoxically, a kind of naïveté, almost an innocence.

Even those evolutionary biologists who admire Wilson's efforts to lay the foundations for a detailed theory of human nature doubt whether such an endeavor can succeed. Dawkins, for example, loathed the "knee-jerk hostility" toward sociobiology displayed by Stephen Jay Gould and other left-leaning scientists. "I think Wilson was shabbily treated, not least by his colleagues at Harvard," Dawkins said. "And so if there's an opportunity to be counted, I would stand up and be counted with Wilson." Yet Dawkins was not as confident as Wilson seemed to be that "the messiness of human life" could be completely understood in scientific terms. Science is not intended to explain "highly complex systems arising out of lots and lots of details," Dawkins elaborated. "Explaining sociology would be rather like using science to explain or to predict the exact course of a molecule of water as it goes over Niagara Falls. You couldn't do it, but that doesn't mean that there's anything *fundamentally* difficult about it. It's just very, very complicated."

I suspect that Wilson himself may doubt whether sociobiology will ever become as all-powerful as he once believed. At the end of *Sociobiology*, he implied that the field would eventually culminate in a complete, final theory of human nature. "To maintain the species indefinitely," Wilson wrote, "we are compelled to drive toward total knowledge, right down to the levels of the neuron and the gene. When we have progressed enough to explain ourselves in these mechanistic terms, and the social sciences come to full flower, the result might be hard to accept." He closed with a quote from Camus: "in a universe divested of illusions and lights, man feels an alien, a stranger. His exile is without remedy since he is deprived of the memory of a lost home or the hope of a promised land."[8]

When I recalled this gloomy coda, Wilson admitted that he had finished *Sociobiology* in a slight depression. "I thought after a period of time, as we knew more and more about where we came from and why we do what we do, in precise terms, that it would reduce—what's the word I'm looking for—our exalted self-image, and our hope for indefinite growth in the future." Wilson also believed that such a theory would bring about the end of biology, the discipline that had given meaning to his own life. "But then I talked myself out of that," he said. Wilson decided that the human mind, which has been and is still being shaped by the complex interaction between culture and genes, represented an endless frontier for science. "I saw that here was an immense unmapped area of science and human history which we would take forever to explore," he recalled. "That made me feel much more cheerful." Wilson resolved his depression by acknowledging, in effect, that his critics were right: science cannot explain all the vagaries of human thought and culture. There can be no *complete* theory of human nature, one that resolves all the questions we have about ourselves.

Just how revolutionary is sociobiology? Not very, according to Wilson himself. For all his creativity and ambition, Wilson is a rather conventional Darwinian. That became clear when I asked him about a concept called biophilia, which holds that the human affinity for nature, or at least certain aspects of it, is innate, a product of natural selection. Biophilia represents Wilson's effort to find common ground between his two great passions, sociobiology and biodiversity. Wilson wrote a monograph on biophilia, published in 1984, and later edited a collection of essays on the topic by himself and others. During my conversation with Wilson, I made the mistake of remarking that biophilia reminded me of Gaia, because each idea evokes an altruism that embraces all of life rather than just one's kin or even one's species.

"Actually, not," Wilson replied, so sharply that I was taken aback. Biophilia does not posit the existence of "some phosphorescent altruism in the air," Wilson scoffed. "I take a very strong mechanistic view of where human nature came from," he said. "Our concern for other organisms is very much a product of Darwinian natural selection." Biophilia evolved, Wilson continued, not for the benefit of all life, but for the benefit of individual humans. "My view is pretty strictly anthropocentric, because what I see and understand, from all that I know of evolution, supports that view and not the other."

I asked Wilson whether he agreed with his Harvard colleague Ernst Mayr that modern biology had been reduced to addressing puzzles whose solution would merely reinforce the prevailing paradigm of neo-Darwinism.[9] Wilson smirked. "Fix the constants to the next decimal point," he said, alluding to the quote that helped to create the legend of the complacent nineteenth-century physicists. "Yeah, we've heard that." But having gently mocked Mayr's view of completion, Wilson went on to agree with it. "We are not about to dethrone evolution by natural selection, or our basic understanding of speciation," Wilson said. "So I, too, am skeptical that we are going to go through any revolutionary changes of how evolution works or how diversification works or how biodiversity is created, at the species level." There is much to learn about embryonic development, about the interaction between human biology and culture, about ecologies and other complex systems. But the basic rules of biology, Wilson asserted, are "beginning to fall pretty much, in my judgment, permanently into place. How evolution works, the algorithm, the machine, what drives it."

What Wilson might have added is that the chilling moral and philosophical implications of Darwinian theory were spelled out long ago. In his 1871 book, *The Descent of Man*, Darwin noted that if humans had evolved as bees had, "there can be hardly a doubt that our unmarried females would, like the worker-bees, think it a sacred duty to kill their brothers, and mothers would strive to kill their fertile daughters; and no one would think of interfering."[10] In other words, we humans are animals, and natural selection has shaped not only our bodies but our very beliefs, our fundamental sense of right and wrong. One dismayed Victorian reviewer of *Descent* fretted in the Edinburgh Review, "If these views be true, a revolution in thought is imminent, which will shake society to its very foundations by destroying the sanctity of the conscience and the religious sense."[11] That revolution happened long ago. Before the end of the nineteenth century, Nietzsche had proclaimed that there were no divine underpinnings to human morality: God is dead. We did not need sociobiology to tell us that.

A Few Words from Noam Chomsky

One of the more intriguing critics of sociobiology and other Darwinian approaches to social science is Noam Chomsky, who is both a linguist and one of America's most uncompromising social critics. The first time I saw Chomsky in the flesh he was giving a talk on the practices of modern labor unions. He was wiry, with the slight hunch of a chronic reader. He wore steel-rimmed glasses, sneakers, chinos, and an open-necked shirt. But for the lines in his face and the gray in his longish hair, he could have passed for a college student, albeit one who would rather discuss Hegel than guzzle beer at fraternity parties.

Chomsky's main message was that union leaders were more concerned with maintaining their own power than with representing workers. His audience? Union leaders. During the question-and-answer period they reacted, as one might expect, with defensiveness and even hostility. But Chomsky met their arguments with such serene, unshakable conviction— and such a relentless barrage of *facts*—that before long the targets of his criticism were nodding in agreement: yes, perhaps they *were* selling out to their capitalist overlords.

When I later expressed surprise to Chomsky at the harshness of his lecture, he informed me that he was not interested in "giving people A's for being right." He opposed all authoritarian systems. Of course, he usually focused his wrath not on labor unions, which have lost much of their power, but on the U.S. government, industry, and the media. He called the United States a "terrorist superpower" and the media its "propaganda agent." He told me that if the *New York Times*, one of his favorite targets, started reviewing his books on politics, it would be a sign to him that he was doing something wrong. He summed up his world view as "whatever the establishment is, I'm against it."

I said I found it ironic that his political views were so antiestablishment, given that in linguistics he *is* the establishment. "No I'm not," he snapped. His voice, which ordinarily is hypnotically calm—even when he is eviscerating someone—suddenly had an edge. "My position in linguistics is a minority position, and it always has been." He insisted that he was "almost totally incapable of learning languages" and that, in fact, he was not even a professional linguist. MIT had only hired him and given him tenure, he suggested, because it really did not know or care much about the humanities; it simply needed to fill a slot.[12]

I provide this background for its cautionary value. Chomsky is one of the most contrarian intellectuals I have met (rivaled only by the anarchic philosopher Paul Feyerabend). He is compelled to put all authority figures in their places, even himself. He exemplifies the anxiety of self-influence. One should thus take all of Chomsky's pronouncements with a grain of salt. In spite of his denials, Chomsky is the most important linguist who has ever lived. "It is hardly an exaggeration to say that there is no major theoretical issue in linguistics today that is debated in terms other than those in which he has chosen to define it," declares the *Encyclopaedia Britannica*.[13] Chomsky's position in the history of ideas has been likened to that of Descartes and Darwin.[14] When Chomsky was in graduate school in the 1950s, linguistics—and all the social sciences—was dominated by behaviorism, which hewed to John Locke's notion that the mind begins as a *tabula rasa*, a blank slate that is inscribed upon by experience. Chomsky challenged this approach. He contended that children could not possibly learn language solely through induction, or trial and error, as behaviorists believed. Some fundamental principles of language—a kind of universal grammar—must be embedded in our brains. Chomsky's theories, which he first set forth in his 1957 book *Syntactic Structures*, helped to rout behaviorism once and for all and paved the way for a more Kantian, genetically oriented view of human language and cognition.[15]

Edward Wilson and other scientists who attempt to explain human nature in genetic terms are all, in a sense, indebted to Chomsky. But Chomsky has never been comfortable with Darwinian accounts of human behavior. He accepts that natural selection may have played *some* role in the evolution of language and other human attributes. But given the enormous gap between human language and the relatively simple communication systems of other animals, and given our fragmentary knowledge of the past, science can tell us little about how language evolved. Just because language is adaptive now, Chomsky elaborates, does not mean that it arose in response to selection pressures. Language may have been an incidental by-product of a spurt in intelligence that only later was coopted for various uses. The same may be true of other properties of the human mind. Darwinian social science, Chomsky has complained, is not a real science at all but "a philosophy of mind with a little bit of science thrown in." The problem, according to Chomsky, is that "Darwinian theory is so loose it can incorporate anything they discover."[16]

Chomsky's evolutionary perspective has, if anything, convinced him that we may have only a limited ability to understand nature, human or inhuman. He rejects the notion—popular among many scientists—that evolution shaped the brain into a general-purpose learning and problem-solving machine. Chomsky believes, as Gunther Stent and Colin McGinn do, that the innate structure of our minds imposes limits on our understanding. (Stent and McGinn arrived at this conclusion in part because of Chomsky's research.)

Chomsky divides scientific questions into problems, which are at least potentially answerable, and mysteries, which are not. Before the seventeenth century, Chomsky explained to me, when science did not really exist in the modern sense, almost all questions appeared to be mysteries. Then Newton, Descartes, and others began posing questions and solving them with the methods that spawned modern science. Some of those investigations have led to "spectacular progress," but many others have proved fruitless. Scientists have made absolutely no progress, for example, investigating such issues as consciousness and free will. "We don't even have bad ideas," Chomsky said.

All animals, he argued, have cognitive abilities shaped by their evolutionary histories. A rat, for example, may learn to navigate a maze that requires it to turn left at every second fork but not one that requires it to turn left at every fork corresponding to a prime number. If one believes that humans are animals—and not "angels," Chomsky added sarcastically—then we, too, are subject to these biological constraints. Our language capacity allows us to formulate questions and resolve them in ways that rats cannot, but ultimately we, too, face mysteries as absolute as that faced by the rat in a prime-number maze. We are limited in our ability to ask questions as well. Chomsky thus rejected the possibility that physicists or other scientists could attain a theory of everything; at best, physicists can only create a "theory of what they know how to formulate."

In his own field of linguistics "there's a lot of understanding now about how human languages are more or less cast in the same mold, what the principles are that unify them and so on." But many of the most profound issues raised by language remain impenetrable. Descartes, for instance, struggled to comprehend the human ability to use language in endlessly creative ways. "We're facing the same blank wall Descartes did" on that issue, Chomsky said.

In his 1988 book, *Language and Problems of Knowledge*, Chomsky suggested that our verbal creativity may prove more fruitful than our scientific skills for addressing many questions about human nature. "It is quite possible—overwhelmingly probable, one might guess—that we will always learn more about human life and human personality from novels than from scientific psychology," he wrote. "The science-forming capacity is only one facet of our mental endowment. We use it where we can but are not restricted to it, fortunately."[17]

The success of science, Chomsky proposed to me, stems from "a kind of chance convergence of the truth about the world and the structure of our cognitive space. And it *is* a chance convergence because evolution didn't design us to do this; there's no pressure on differential reproduction that led to the capacity to solve problems in quantum theory. We had it. It's just there for the same reason that most other things are there: for some reason that nobody understands."

Modern science has stretched the cognitive capacity of humans to the breaking point, according to Chomsky. In the nineteenth century, any well-educated person could grasp contemporary physics, but in the twentieth century "you've got to be some kind of freak." That was my opening. Does the increasing difficulty of science, I asked, imply that science might be approaching its limits? Might science, defined as the search for *comprehensible* regularities or patterns in nature, be ending? Abruptly, Chomsky took back everything he had just said. "Science is hard, I would agree with that. But when you talk to young children, they want to understand nature. It's driven out of them. It's driven out of them by boring teaching and by an educational system that tells them they're too stupid to do it." Suddenly it was the establishment, not our innate limitations, that had brought science to its current impasse.

Chomsky insisted that "there are major questions for the natural sciences which we can formulate and that are within our grasp, and that's an exciting prospect." For example, scientists still must show—and almost certainly will show—how fertilized cells grow into complex organisms and how the human brain generates language. There is still plenty of science left to do, "plenty of physics, plenty of biology, plenty of chemistry."

In denying the implication of his own ideas, Chomsky may have been exhibiting just another odd spasm of self-defiance. But I suspect he was really succumbing to wishful thinking. Like so many other scientists, he cannot imagine a world without science. I once asked Chomsky which work he found more satisfying, his political activism or his linguistic research. He seemed surprised that I needed to ask. Obviously, he replied, he spoke out against injustice merely out of a sense of duty; he took no intellectual pleasure from it. If the world's problems were suddenly to disappear, he would happily, joyfully, devote himself to the pursuit of knowledge for its own sake.

The Antiprogress of Clifford Geertz

Practitioners of ironic science can be divided into two types: naïfs, who believe or at least hope they are discovering objective truths about nature (the superstring theorist Edward Witten is the archetypal example), and sophisticates, who realize that they are, in fact, practicing something more akin to art or literary criticism than to conventional science. There is no better example of a sophisticated ironic scientist than the anthropologist Clifford Geertz. He is simultaneously a scientist and a philosopher of science; his work is one long comment on itself. If Stephen Jay Gould serves as the negative capability of evolutionary biology, Geertz does the same for social science. Geertz has helped to fulfill Gunther Stent's prophecy in *The Coming of the Golden Age* that the social sciences "may long remain the ambiguous, impressionistic disciplines that they are at present."[18]

I first encountered Geertz's writings in college, when I took a class on literary criticism and the instructor assigned Geertz's 1973 essay "Thick Description: Toward an Interpretive Theory of Culture."[19] The essay's basic message was that an anthropologist cannot portray a culture by merely "recording the facts." He or she must interpret phenomena, must try to guess what they *mean*. Consider the blink of an eye, Geertz wrote (crediting the example to the British philosopher Gilbert Ryle). The blink may represent an involuntary twitch stemming from a neurological disorder, or from fatigue, or from nervousness. Or it may be a wink, an intentional signal, with many possible meanings. A culture consists of a virtually infinite number of such messages, or signs, and the anthropologist's task is to interpret them. Ideally, the anthropologist's interpretation of a culture should be as complex and richly imagined as the culture itself. But just as literary critics never can hope to establish, once and for all, what *Hamlet* means, so anthropologists must eschew all hope of discovering absolute truths. "Anthropology, or at least interpretive anthropology, is a science whose progress is marked less by a perfection of consensus than by a refinement of debate," Geertz wrote. "What gets better is the precision with which we vex each other."[20] The point of his brand of science, Geertz realized, is not to bring discourse to a close, but to perpetuate it in ever-more-interesting ways.

In later writings, Geertz likened anthropology not only to literary criticism but also to literature. Ethnography involves "telling stories, making pictures, concocting symbolisms and deploying tropes," Geertz wrote, just as literature does. He called anthropology "faction," or "imaginative writing about real people in real places at real times."[21] (Of course, substituting art for literary criticism hardly represents a radical step for one such as Geertz, since to most postmodernists a text is a text is a text.)

Geertz displayed his own talents as a faction writer in "Deep Play: Notes on the Balinese Cockfight." The first sentence of the 1972 essay established his anything-but-straightforward style: "Early in April of 1958, my wife and I arrived, malarial and diffident, in a Balinese village we intended, as anthropologists, to study."[22] (Geertz's prose has been likened to that of both Marcel Proust and Henry James. Geertz told me he was flattered by the former comparison, but thought the latter was probably closer to the mark.)

The opening section of the essay described how the young couple gained the confidence of the normally aloof Balinese. Geertz, his wife, and a group of villagers had been watching a cockfight when police raided the scene. The American couple fled along with their Balinese neighbors. Impressed that the scientists had not sought privileged treatment from the police, the villagers accepted them.

Having thus established his credentials as an insider, Geertz proceeded to depict and then to analyze the Balinese obsession with cockfighting. He eventually concluded that the bloody sport—in which roosters armed with razor-sharp spurs fight to the death—mirrored and thus exorcized the Balinese people's fear of the dark forces underlying their superficially calm society. Like *King Lear* or *Crime and Punishment*, cockfighting "catches up these themes—death, masculinity, rage, pride, loss, beneficence, chance— and [orders] them into an encompassing structure."[23]

Geertz is a shambling bear of a man, with shaggy, whitening hair and beard. When I first interviewed him one drizzly spring day at the Institute for Advanced Study in Princeton, he fidgeted incessantly, pulling on his ear, pawing his cheek, slouching down in his chair and abruptly drawing himself up.[24] Now and then, as he listened to me pose a question, he pulled the top of his sweater up over the tip of his nose, like a bandit trying to conceal his identity. His discourse, too, was elusive. It mimicked his writing: all stops and starts, headlong assertions punctuated by countless qualifications, and suffused with hypertrophic self-awareness.

Geertz was determined to correct what he felt was a common mis-impression, that he was a universal skeptic who did not believe that science could achieve any durable truths. Some fields, Geertz said, notably physics, obviously do have the capability to arrive at the truth. He also stressed that, contrary to what I might have heard, he did not consider anthropology to be merely an art form, devoid of any empirical content, and thus not a legitimate field of science. Anthropology is "empirical, responsive to evidence, it theorizes," Geertz said, and practitioners can sometimes achieve a nonabsolute falsification of ideas. Hence it is a science, one that can achieve progress of a kind.

On the other hand, "nothing in anthropology has anything like the status of the harder parts of the hard sciences, and I don't think it ever will," Geertz said. "Some of the assumptions that [anthropologists] made about how easy it is to understand all this and what you need to do in order to do that are no longer really—well, nobody believes them anymore." He laughed. "It doesn't mean it's impossible to know anybody, either, or to do anthropological work. I don't think that at all. But it's not easy."

In modern anthropology, disagreement rather than consensus is the norm. "Things get more and more complicated, but they don't converge to a single point. They spread out and disperse in a very complex way. So I don't see everything heading toward some grand integration. I see it as much more pluralistic and differentiated."

As Geertz continued speaking, it seemed that the progress he envisioned was a kind of antiprogress, in which anthropologists would eliminate, one by one, all the assumptions that could make consensus possible; firm beliefs would dwindle, and doubts multiply. He noted that few anthropologists still believed that they could extract universal truths about humanity from the study of so-called "primitive" tribes, those supposedly existing in a pristine state, uncorrupted by modern culture; neither could anthropologists pretend to be purely objective data gatherers, free of biases and preconceptions.

Geertz found laughable the predictions of Edward Wilson that the social sciences could eventually be rendered as rigorous as physics by grounding them in evolutionary theory, genetics, and neuroscience. Would-be revolutionaries have always come forward with some grand idea that would unify the social sciences, Geertz recalled. Before sociobiology there was general systems theory, and cybernetics, and Marxism. "The notion that someone is going to come along and revolutionize everything overnight is a kind of academicians' disease," Geertz said.

At the Institute for Advanced Study, Geertz was occasionally approached by physicists or mathematicians who had developed highly mathematical models of racial relations and other sociological problems. "But they don't know anything about what goes on in the inner cities!" Geertz exclaimed. "They just have a mathematical model!" Physicists, he grumbled, would never stand for a theory of physics that lacked an empirical foundation. "But somehow or other social science doesn't count. And if you want to have a general theory of war and peace, all you have to do is to sit down and write an equation without having any knowledge of history or people."

Geertz was painfully aware that the introspective, literary style of science he had promulgated also had its pitfalls. It could lead to excessive self-consciousness, or "epistemological hypochondria" on the part of the practitioner. This trend, which Geertz dubbed "I-witnessing," had produced some interesting works but also some abysmal ones. Some anthropologists, Geertz noted, had become so determined to expose all their potential biases, ideological or otherwise, that their writings resembled confessionals, revealing far more about the author than about the putative subject.

Geertz had recently revisited two regions he had studied early in his career, one in Morocco and the other in Indonesia. Both places had changed, drastically; he had also changed. As a result, he became even *more* aware of just how hard it is for the anthropologist to discern truths that transcend their time, place, context. "I always felt it could end in total failure," he said. "I'm still reasonably optimistic, in the sense that I think it's doable, doable as long as you don't claim too much for it. Am I pessimistic? No, but I am chastened." Anthropology is not the only field grappling with questions about its own limitations, Geertz pointed out. "I do sense the same mood in all kinds of fields"—even in particle physics, he said, which seems to be reaching the limits of empirical verification. "The kind of simple self-confidence in science that there once was doesn't seem to me to be so pervasive. Which doesn't mean everybody is giving up hope and wringing their hands in anguish and so on. But it is extraordinarily difficult."

At the time of our meeting in Princeton, Geertz was writing a book about his excursions into his past. When the book was published in 1995, its title neatly summarized Geertz's anxious attitude: *After the Fact*. Geertz peeled apart the title's multiple meanings in the book's final paragraph: Scientists like him are, of course, chasing after facts, but they can only capture facts, if at all, retrospectively; by the time they reach some understanding of what has taken place, the world has moved on, inscrutable as ever.

The phrase also referred, Geertz concluded, to the "post-positivist critique of empirical realism, the move away from simple correspondence theories of truth and knowledge which makes of the very term 'fact' a delicate matter. There is not much assurance or sense of closure, not even much of a sense of knowing what it is one precisely *is* after, in so indefinite a quest, amid such various people, over such a diversity of times. But it is an excellent way, interesting, dismaying, useful, and amusing, to spend a life."[25] Ironic social science may not get us anywhere, but at least it can give us something to do, forever if we like.

CHAPTER SEVEN

The End of Neuroscience

Mind, not space, is science's final frontier. Even the most avid believers in the power of science to solve its problems consider the mind a potentially endless source of questions. The problem of mind can be approached in many ways. There is the historical dimension: How, why, did *Homo sapiens* become so smart? Darwin provided a general answer long ago: natural selection favored hominids who could use tools, anticipate the actions of potential competitors, organize into hunting parties, share information through language, and adapt to changing circumstances. Together with modern genetics, Darwinian theory has much to say about the structure of our minds and thus about our sexual and social behavior (although not as much as Edward Wilson and other sociobiologists might like).

But modern neuroscientists are interested less in how and why our minds evolved, in a historical sense, than in how they are structured and work right now. The distinction is similar to the one that can be made between cosmology, which seeks to explain the origins and subsequent evolution of matter, and particle physics, which addresses the structure of matter as we find it here in the present. One discipline is historical and thus necessarily tentative, speculative, and open-ended. The other is, by comparison, much more empirical, precise, and amenable to resolution and finality.

Even if neuroscientists restricted their studies to the mature rather than embryonic brain, the questions would be legion. How do we learn, remember, see, smell, taste, and hear? Most researchers would say these problems, although profoundly difficult, are tractable; scientists will solve them by reverse engineering our neural circuitry. Consciousness, our subjective sense of awareness, has always seemed to be a different sort of puzzle, not physical but metaphysical. Through most of this century, consciousness was not considered a proper subject for scientific investigation. Although behaviorism had died, its legacy lived on in the reluctance of scientists to consider subjective phenomena, and consciousness in particular.

That attitude changed when Francis Crick turned his attention to the problem. Crick is one of the most ruthless reductionists in the history of science. After he and James Watson unraveled the twin-corkscrew structure of DNA in 1953, Crick went on to show how genetic information is encoded into DNA. These achievements gave Darwin's theory of evolution and Mendel's theory of heredity the hard empirical base they had lacked. In the mid-1970s, Crick moved from Cambridge, England, where he had spent most of his career, to the Salk Institute for Biological Research, a cubist fortress overlooking the Pacific Ocean just north of San Diego, California. He worked on developmental biology and the origin of life before finally turning his attention to the most elusive and inescapable of all phenomena: consciousness. Only Nixon could go to China. And only Francis Crick could make consciousness a legitimate subject for science.[1]

In 1990, Crick and a young collaborator, Christof Koch, a German-born neuroscientist at the California Institute of Technology, proclaimed in *Seminars in the Neurosciences* that it was time to make consciousness a subject of empirical investigation. They asserted that one could not hope to achieve true understanding of consciousness or any other mental phenomena by treating the brain as a black box, that is, an object whose internal structure is unknown and even irrelevant. Only by examining neurons and the interactions between them could scientists accumulate the kind of unambiguous knowledge needed to create truly scientific models of consciousness, models analogous to those that explain heredity in terms of DNA.[2]

Crick and Koch rejected the belief of many of their colleagues that consciousness could not be defined, let alone studied. Consciousness, they argued, was synonymous with awareness, and all forms of awareness—whether directed toward objects in the world or highly abstract, internal concepts—seem to involve the same underlying mechanism, one that combines attention with short-term memory. (Crick and Koch credited William James with inventing this definition.) Crick and Koch urged investigators to focus on visual awareness as a synecdoche for consciousness, since the visual system had been so well mapped. If researchers could find the neural mechanisms underlying this function, they might be able to unravel more complex and subtle phenomena, such as self-consciousness, that might be unique to humans (and thus much more difficult to study at the neural level). Crick and Koch had done the seemingly impossible: they had transformed consciousness from a philosophical mystery to an empirical problem. A theory of consciousness would represent the apogee—the culmination—of neuroscience.

Legend has it that some students of the arch-behaviorist B. F. Skinner, on being exposed firsthand to his relentlessly mechanistic view of human nature, fell into an existential despair. I recalled this factoid on meeting Crick in his huge, airy office at the Salk Institute. He was not a gloomy or dour man. Quite the contrary. Clad in sandals, corn-colored slacks, and a gaudy Hawaiian shirt, he was almost preternaturally jolly. His eyes and mouth curled up at the corners in a perpetual wicked grin. His bushy white eyebrows flared out and up like horns. His ruddy face flushed even darker when he laughed, which he did often and with gusto. Crick seemed particularly cheery when he was skewering some product of wishful and fuzzy thinking, such as my vain hope that we humans have free will.[3]

Even an act as apparently simple as seeing, Crick informed me in his crisp, Henry Higgins accent, actually involves vast amounts of neural activity. "The same could be said as to how you make a move, say, picking up a pen," he continued, plucking a ballpoint from his desk and waving it before me. "A lot of computation goes on preparing you for that movement. What you're aware of is a decision, but you're not aware of what makes you do the decision. It seems free to you, but it's the result of things you're not aware of." I frowned, and Crick chuckled.

Trying to help me understand what he and Koch meant by attention—which was the crucial component of their definition of consciousness—Crick emphasized that it involved more than the simple processing of information. To demonstrate this point, he handed me a sheet of paper imprinted with a familiar black-and-white pattern: I saw a white vase on a black background one moment and two silhouetted human profiles the next. Although the visual input to my brain remained constant, Crick pointed out, what I was aware of—or paying attention to—kept changing. What change in the brain corresponded to this change in attention? If neuroscientists could answer that question, Crick said, they might go far toward solving the mystery of consciousness.

Crick and Koch had actually offered a tentative answer to this question in their 1990 paper on consciousness. Their hypothesis was based on evidence that when the visual cortex is responding to stimulation, certain groups of neurons fire extremely rapidly and in synchrony. These oscillating neurons, Crick explained to me, might correspond to aspects of the scene to which attention is being directed. If one envisions the brain as a vast, muttering crowd of neurons, the oscillating neurons are like a group of people who suddenly start singing the same song. Going back to the vase-profiles figure, one set of neurons sings "vase" and then another sings "faces."

The oscillation theory (which has been advanced independently by other neuroscientists) has its weaknesses, as Crick was quick to admit. "I think it's a good, brave, first attempt," he said, "but I have my doubts that it will turn out to be right." He observed that he and Watson had only succeeded in discovering the double helix after numerous false starts. "Exploratory research is really like working in a fog. You don't know where you're going. You're just groping. Then people learn about it afterwards and think how straightforward it was." Crick was nonetheless confident the issue would be resolved not by arguing over psychological concepts and definitions, but by doing "lots of experiments. That's what science is about."

Neurons must be the basis for any model of the mind, Crick told me. Psychologists have treated the brain as a black box, which can be understood in terms merely of inputs and outputs rather than of internal mechanisms. "Well, you can do that if the black box is simple enough, but if the black box is complicated the chances of you getting the right answer are rather small," Crick said. "It's just the same as in genetics. We had to know about genes, and what genes did. But to pin it down, we had to get down to the nitty-gritty and find the molecules and things involved."

Crick gloated that he was in a perfect position to promote consciousness as a scientific problem. "I don't have to get grants," he said, because he had an endowed chair at the Salk Institute. "The main reason I do it is because I find the problem fascinating, and I feel I've earned the right to do what I like." Crick did not expect researchers to solve these problems overnight.

"What I want to stress is that the problem is important and has been too long neglected."

While talking to Crick, I could not help but think of the famous first line of *The Double Helix,* James Watson's memoir about how he and Crick had deciphered DNA's structure: "I have never seen Francis Crick in a modest mood."[4] Some historical revisionism is in order here. Crick is often modest. During our conversation, he expressed doubts about his oscillation theory of consciousness; he said parts of a book he was writing on the brain were "dreadful" and needed rewriting. When I asked Crick how he interpreted Watson's quip, he laughed. What Watson meant, Crick suggested, was not that he was immodest, but that he was "full of confidence and enthusiasm and things like that." If he was also a tad bumptious at times, and critical of others, well, that was because he wanted so badly to get to the bottom of things. "I can be patient for about 20 minutes," he said, "but that's it."

Crick's analysis of himself, like his analysis of most things, seemed on target. He has the perfect personality for a scientist, an empirical scientist, the kind who answers questions, who gets us somewhere. He is, or appears to be, singularly free of self-doubt, wishful thinking, and attachments to his own theories. His immodesty, such as it is, comes simply from wanting to know how things work, regardless of the consequences. He cannot tolerate obfuscation or wishful thinking or untestable speculation, the hallmarks of ironic science. He is also eager to share his knowledge, to make things as clear as possible. This trait is not as common among prominent scientists as one might expect.

In his autobiography, Crick revealed that as a youth and would-be scientist he had worried that by the time he grew up everything would already be discovered. "I confided my fears to my mother, who reassured me. 'Don't worry, Ducky,' she said. 'There will be plenty left for you to find out.' "[5] Recalling this passage, I asked Crick if he thought there would *always* be plenty left for scientists to find out. It all depends on how one defines science, he replied. Physicists might soon determine the fundamental rules of nature, but they could then apply that knowledge by inventing new things forever. Biology seems to have an even longer future. Some biological structures—such as the brain—are so complex that they may resist elucidation for some time. Other puzzles, especially historical ones, such as the origin of life, may never be fully answered simply because the available data are insufficient. "There are enormous numbers of interesting problems" in biology, Crick said. "There's enough to keep us busy at least through our grandchildren's time." On the other hand, Crick agreed with Richard Dawkins that biologists already had a good general understanding of the processes underlying evolution.

As Crick escorted me out of his office, we passed a table bearing a thick stack of paper. It was a draft of Crick's book on the brain, entitled *The Astonishing Hypothesis*. Would I like to read his opening paragraph? Sure, I said. "The Astonishing Hypothesis," the book began, "is that 'You,' your joys and your sorrows, your memories and your ambitions, your sense of personal identity and free will, are in fact no more than the behavior of a vast assembly of nerve cells and their associated molecules. As Lewis Caroll's Alice might have phrased it, 'You're nothing but a pack of neurons.' "[6] I looked at Crick. He was grinning from ear to ear.

I was talking to Crick on the telephone several weeks later, checking the facts of an article I had written about him, when he asked me for some advice. He confessed that his editor wasn't thrilled with the title *The Astonishing Hypothesis*; she didn't think the view that "we are nothing but a pack of neurons" was all that astonishing. What did I think? I told Crick I had to agree; his view of the mind was, after all, just old-fashioned reductionism and materialism. I suggested that *The Depressing Hypothesis* might be a more fitting title, but it might repel would-be readers. The title didn't matter that much anyway, I added, since the book would sell on the strength of Crick's name.

Crick absorbed all this with his usual good humor. When his book appeared in 1994, it was still called *The Astonishing Hypothesis*. However, Crick, or, more likely, his editor, had added a subtitle: *The Scientific Search for the Soul*. I had to smile when I saw it; Crick was obviously trying not to find the soul—that is, some spiritual essence that exists independently of our fleshy selves—but to eliminate the possibility that there was one. His DNA discovery had gone far toward eradicating vitalism, and now he hoped to stamp out any last vestiges of that romantic worldview through his work on consciousness.

Gerald Edelman Postures around the Riddle

One of the premises of Crick's approach to consciousness is that no theory of mind advanced to date has much value. Yet at least one prominent scientist, a Nobel laureate no less, claims to have gone far toward solving the problem of consciousness: Gerald Edelman. His career, like Crick's, has been eclectic, and highly successful. While still a graduate student, Edelman helped to determine the structure of immunoglobulins, proteins that are crucial to the body's immune response. In 1972 he shared a Nobel Prize for that work. Edelman moved on to developmental biology, the study of how a single fertilized cell becomes a full-fledged organism. He found a class of proteins, called cell-adhesion molecules, thought to play an important role in embryonic development.

All this was merely prelude, however, to Edelman's grand project of creating a theory of mind. Edelman has set forth his theory in four books: *Neural Darwinism, Topobiology, The Remembered Present*, and *Bright Air, Brilliant Fire.*[7] The gist of the theory is that just as environmental stresses select the fittest members of a species, so inputs to the brain select groups of neurons—corresponding to useful memories, for example—by strengthening the connections between them.

Edelman's tumescent ambition and personality have made him an alluring subject for journalists. A *New Yorker* profile called him "a dervish of motion, energy and raw intellect" who was "as much Henny Youngman as Einstein"; it mentioned that detractors consider him "an empire-building egomaniac."[8] In a *New York Times Magazine* cover story in 1988, Edelman conferred divine powers upon himself. Discussing his work in immunology, he said that "before I came to it, there was darkness—afterwards there was light." He called a robot based on his neural model his "creature" and said: "I can only observe it, like God. I look down on its world."[9]

I experienced Edelman's self-regard firsthand when I visited him at Rockefeller University in June 1992. (Sometime later, Edelman left Rockefeller to head his own laboratory at the Scripps Institute in La Jolla, California, just down the road from Crick.) Edelman is a large man. Clad in a dark, broad-shouldered suit, he exuded a kind of menacing elegance and geniality. As in his books, he kept interrupting his scientific discourse to dispense stories, jokes, or aphorisms whose relevance was often obscure.

The digressions seemed intended to demonstrate that Edelman represented the ideal intellectual, both cerebral and earthy, learned and worldly; no mere experimentalist, he.

Explaining how he became interested in the mind, Edelman said: "I'm very excited by dark and romantic and open issues of science. I'm not averse to working on details, but pretty much only in the service of trying to address this issue of closure." Edelman wanted to find the answer to great questions. His Nobel Prize–winning research on antibody structure had transformed immunology into "more or less a closed science"; the central question, which concerned how the immune system responds to invaders, was resolved. He and others helped to show that self-recognition happens through a process known as selection: the immune system has innumerable different antibodies, and the presence of foreign antigens spurs the body to accelerate the production of, or select, antibodies specific to that antigen and to suppress the production of other antibodies.

Edelman's search for open questions led him inexorably to the development and operation of the brain. He realized that a theory of the human mind would represent the ultimate closure for science, for then science could account for its own origin. Consider superstring theory, Edelman said. Could it explain the existence of Edward Witten? Obviously not. Most theories of physics relegate issues involving the mind to "philosophy or sheer speculation," Edelman noted. "You read that section of my book where Max Planck says we'll never get this mystery of the universe because we are the mystery? And Woody Allen said if I had my life to live over again I'd live in a delicatessen?"

Describing his approach to the mind, Edelman sounded, at first, as resolutely empirical as Crick. The mind, Edelman emphasized, can only be understood from a biological standpoint, not through physics or computer science or other approaches that ignore the structure of the brain. "We will not have a deeply satisfactory brain theory unless we have a deeply satisfactory theory of neural anatomy, okay? It's as simple as that." To be sure, "functionalists," such as the artificial-intelligence maven Marvin Minsky, say they can build an intelligent being without paying attention to anatomy. "My answer is, 'When you show me, fine.' "

But as Edelman continued speaking, it became clear that, unlike Crick, he viewed the brain through the filter of his idiosyncratic obsessions and ambitions. He seemed to think that all his insights were totally original; no one had truly seen the brain before he had turned his attention to it. He recalled that when he had started studying the brain, or, rather, brains, he was immediately struck by their variability. "It seemed to me very curious that the people who worked in neuroscience always talked about brains as if they were identical," he said. "When you look at papers everybody talked about it as if it were a replicable machine. But when you actually look in depth, at every level—and there are amazing numbers of levels—the thing that really hits you is the diversity." Even identical twins, he remarked, show great differences in the organization of their neurons. These differences, far from being insignificant noise, are profoundly important. "It's quite scary," Edelman said. "That's something you just can't get around."

The vast variability and complexity of the brain may be related to a problem with which philosophers from Kant to Wittgenstein had wrestled: how do we categorize things? Wittgenstein, Edelman elaborated, high-lighted the troublesome nature of categories by pointing out that different games often had nothing in common except the fact that they were games. "Typical Wittgenstein," Edelman mused. "There is a kind of ostentation in his modesty. I don't know what that is. He provokes you and it's very powerful. It's ambiguous, sometimes, and it's not cute. It's riddle, it's posturing around the riddle."

A little girl playing hopscotch, chess players, and Swedish sailors doing naval exercises are all playing games, Edelman continued. To most ob-servers, these phenomena seem to have little or nothing to do with each other, and yet they are all members of the set of possible games. "This defines what is known in the business as a polymorphous set. It's a very hard thing. It means a set defined by neither necessary nor sufficient conditions. I can show you pictures of it in *Neural Darwinism*." Edelman grabbed the book off his table and flipped through it until he found an illustration of two sets of geometrical forms that represented polymorphous sets. He then pushed the book away and transfixed me once again. "I'm *astonished* that people don't sit and put these things together," Edelman said.

Edelman, of course, did put these things together: the polymorphous diversity of the brain allows it to respond to the polymorphous diversity of nature. The brain's diversity is not irrelevant noise, but is "the very basis on which selection is going to be made, when encountering an unknown set of physical correspondences in the world. All right? Well that's very promis-ing. Let's go one step further. Could the *unit* of selection be the neuron?"

No, because the neuron is too binary, inflexible; it is either on, firing, or off, dormant. But *groups* of linked, interacting neurons could do the job. These groups compete with each other in an effort to create effective representations, or maps, of the infinite variety of stimuli entering from the world. Groups that form successful maps grow still stronger, while other groups wither.

Edelman continued asking and answering his own questions. He spoke slowly, portentously, as if trying to physically impress his words on my brain. How do these groups of linked neurons solve the problem of polymorphous sets that troubled Wittgenstein? Through reentry. What is reentry? "Reentry is the ongoing recursive signaling between mapped areas," Edelman said, "so that you map maps by massively parallel reciprocal connections. It's *not* feedback, which is between two wires, in which I have a definite function, instruction—sine wave in, amplified sine wave out." He was grim, almost angry, as if I had suddenly become the symbol of all his puny-minded, envious critics, who said that reentry *was* merely feedback.

He paused a moment, as if to collect himself, and began again, speaking loudly, slowly, pausing between words, like a tourist trying to make a presumably dim native understand him. Contrary to what his critics said, his model was unique; it had nothing in common with neural networks, he said, lacing the term *neural networks* with scorn. To gain his trust—and because it was true—I confessed that I had always found neural networks difficult to grasp. (Neural networks consist of artificial neurons, or switches, linked by connections of varying strengths.) Edelman smiled triumphantly. "Neural nets involve the stretching of a metaphor," he said. "There is this yawning gap, and you say, 'Is it me, or am I missing something?'" *His* model, he assured me, did not suffer from that problem.

I began asking another question about reentry, but Edelman held up his hand. It was time, he said, to tell me about his latest creature, *Darwin 4.* The best way to validate his theory would be to observe the behavior of neurons in a living animal, which was of course impossible. The only solution, Edelman said, was to construct an automaton that embodied the principles of neural Darwinism. Edelman and his coworkers had built four robots, each named *Darwin*, each more sophisticated than the last. Indeed, *Darwin 4*, Edelman assured me, was not a robot at all but a "real creature." It was "the first nonliving thing that truly learns, okay?"

Again he paused, and I felt his evangelical fervor washing over me. He seemed to be trying to build a sense of drama, as if he were pulling back a succession of veils, each of which concealed a deeper mystery. "Let's go take a look," he said. We headed out of his office and down the hall. He opened the door to a room containing a huge, humming mainframe computer. This, Edelman assured me, was *Darwin 4*'s "brain." Then we walked to another room, where the creature itself awaited us. A pile of machinery on wheels, it sat on a plywood stage littered with blue and red blocks. Perhaps sensing my disappointment—real robots will always disappoint anyone who has seen *Star Wars*—Edelman reiterated that *Darwin 4* "looks like a robot but it's not."

Edelman pointed out the "snout," a bar tipped with a light-sensitive sensor and a magnetic gripper. A television monitor mounted on one wall flashed some patterns that represented, Edelman informed me, the state of *Darwin*'s brain. "When it does find an object it will poke up to it, it'll grab it, and then it will get good or bad values. . . . That will alter the diffuse relationships and synaptology of these things, which are brain maps"—he pointed at the television monitor—"that weaken or strengthen synapses that alter how muscles move."

Edelman stared at *Darwin 4*, which remained stubbornly immobile. "Uh, it takes a fair amount of time," he said, adding that "the amount of computation involved is hair-raising." Finally the robot stirred, to Edelman's evident relief, and began rolling slowly around the platform, nudging blocks, leaving blue ones, picking up red ones with its magnetic snout and taking them over to a big box that Edelman called "home."

Edelman gave me a running commentary. "Uh oh, it just moved its eye. It just found an object. It picked up an object. Now it's going to search for home."

What is its end goal? I asked. "It *has* no end goals," Edelman reminded me with a frown. "We have given it *values*. Blue is bad, red is good." Values are general and thus better suited to helping us cope with the polymorphous world than are goals, which are much more specific. When he was a teenager, Edelman elaborated, he desired Marilyn Monroe, but Marilyn Monroe was not his *goal*. He possessed *values* that led him to desire certain feminine properties, which Marilyn Monroe happened to exemplify.

Brutally repressing an upwelling image of Edelman and Marilyn Monroe, I asked how this robot differed from all the others built by scientists over the past few decades, many of which were capable of feats at least as impressive as those achieved by *Darwin 4*. The difference, Edelman replied, his jaw setting, was that *Darwin 4* possessed values or instincts, whereas other robots needed specific instructions to accomplish any task. But don't all neural networks, I asked, eschew specific instructions for general learning programs? Edelman frowned. "But all of those, you have to exclusively define the input and the output. That's the big distinction. Isn't this correct, Julio?" He turned to a dour, young postdoc who had joined us and was listening to our conversation silently.

After a moment's hesitation, Julio nodded. Edelman, with a broad smile, noted that most artificial-intelligence designers tried to program knowledge in from the top down with explicit instructions for every situation, instead of having knowledge arise naturally from values. Take a dog, he said. Hunting dogs acquire their knowledge from a few basic instincts. "That is more efficacious than any bunch of Harvard boys writing a program for swamps!" Edelman guffawed and glanced at Julio, who joined in uneasily.

But *Darwin 4* is still a computer, a robot, with a limited repertoire of responses to the world, I persisted; Edelman was using language metaphorically when he called it a "creature" with a "brain." As I spoke, Edelman muttered, "Yup, all right, all right," while nodding rapidly. If a computer, he said, is defined as something driven by algorithms, or effective procedures, then *Darwin 4* is *not* a computer. True, computer scientists might program robots to do what *Darwin 4* does. But they would just be faking biological behavior, whereas *Darwin 4*'s behavior is authentically biological. If some random electronic glitch scrambles a line of code in his creature, Edelman informed me, "it'll just correct like a wounded organism and it'll go around again. I do that for the other one and it'll drop dead in its tracks."

Rather than pointing out that all neural networks and many conventional computer programs have this capability, I asked Edelman about the complaints of some scientists that they simply did not understand his theories. Most genuinely new scientific concepts, he replied, must overcome such resistance. He had invited those who had accused him of obscurity—notably Gunther Stent, whose complaints about Edelman's incomprehensibility were quoted in the *New York Times Magazine*—to visit him so he could explain his work in person. (Stent had reached his decision about Edelman's work after sitting next to him during a trans-Atlantic flight.) No one had accepted Edelman's offer. "The opacity, I believe, is in the reception, not the transmission," Edelman said.

By this time, Edelman was no longer making an attempt to hide his irritation. When I asked about his relationship with Francis Crick, Edelman abruptly announced that he had to attend an important meeting. He would leave me in the capable hands of Julio. "I have a very long-term relationship with Francis, and that's not something one can answer—Boom! Boom!—on the way out the door. Or, as Groucho Marx said, 'Leave, and never darken my towels again!' " With that, he departed on a wave of hollow laughter.

Edelman has admirers, but most dwell on the fringes of neuroscience. His most prominent fan is the neurologist Oliver Sacks, whose beautifully written accounts of his dealings with brain-damaged patients have set the standard for literary—that is, ironic—neuroscience. Francis Crick spoke for many of his fellow neuroscientists when he accused Edelman of hiding "presentable" but not terribly original ideas behind a "smoke screen of jargon." Edelman's Darwinian terminology, Crick added, has less to do with any real analogies to Darwinian evolution than with rhetorical grandiosity. Crick suggested that Edelman's theory be renamed "neural Edelmanism." "The trouble with Jerry," Crick said, is that "he tends to produce slogans and sort of waves them about without really paying attention to what other people are saying. So it's really too much hype, is what one is complaining about."[10]

The philosopher Daniel Dennett of Tufts University remained unimpressed after visting Edelman's laboratory. In a review of Edelman's *Bright Air, Brilliant Fire*, Dennett argued that Edelman had merely presented rather crude versions of old ideas. Edelman's denials nothwithstanding, his model *was* a neural network, and reentry *was* feedback, according to Dennett. Edelman also "misunderstands the philosophical issues he addresses at an elementary level," Dennett asserted. Edelman may profess scorn for those who think the brain is a computer, but his use of a robot to "prove" his theory shows that he holds the same belief, Dennett explained.[11]

Some critics accuse Edelman of deliberately trying to take credit for others' ideas by wrapping them in his own idiosyncratic terminology. My own, somewhat more charitable, interpretation is that Edelman has the brain of an empiricist and the heart of a romantic. He seemed to acknowledge as much, in his typically oblique way, when I asked him if he thought science was in principle finite or infinite. "I don't know what that *means*," he replied. "I know what it means when I say that a series in mathematics is finite or infinite. But I don't know what it means to say that science is infinite. Example, okay? I'll quote Wallace Stevens: *Opus Posthumus*. 'In the very long run even the truth doesn't matter. The risk is taken.' " The *search* for truth is what counts, Edelman seemed to be implying, not the truth itself.

Edelman added that Einstein, when asked whether science was exhausted, reportedly answered, "Possibly, but what's the use of describing a Beethoven symphony in terms of air-pressure waves?" Einstein, Edelman explained, was referring to the fact that physics alone could not address questions related to value, meaning, and other subjective phenomena. One might respond by asking: What is the use of describing a Beethoven symphony in terms of reentrant neural loops? How does the substitution of neurons for air-pressure waves or atoms or any physical phenomenon do justice to the magic and mystery of the mind? Edelman cannot accept, as Francis Crick does, that we are "nothing but a pack of neurons." Edelman therefore obfuscates his basic neural theory—infusing it with terms and concepts borrowed from evolutionary biology, immunology, and philosophy—to lend it added grandeur, resonance, mystique. He is like a novelist who risks obscurity—even seeks it—in the hope of achieving a deeper truth. He is a practitioner of ironic neuroscience, one who, unfortunately, lacks the requisite rhetorical skills.

Quantum Dualism

There is one issue on which Crick, Edelman, and indeed almost all neuroscientists agree: the properties of the mind do not depend in any crucial way on quantum mechanics. Physicists, philosophers, and others have speculated about links between quantum mechanics and consciousness since at least the 1930s, when some philosophically inclined physicists began to argue that the act of measurement—and hence consciousness itself—played a vital role in determining the outcome of experiments involving quantum effects. Such theories have involved little more than hand waving, and proponents invariably have ulterior philosophical or even religious motives. Crick's partner Christof Koch summed up the quantum-consciousness thesis in a syllogism: Quantum mechanics is mysterious, and consciousness is mysterious. Q.E.D.: quantum mechanics and consciousness must be related.[12]

One vigorous advocate of a quantum theory of consciousness is John Eccles, a British neuroscientist who won a Nobel Prize in 1963 for his studies of neural transmission. Eccles is perhaps the most prominent modern scientist to espouse dualism, which holds that the mind exists independently of its physical substrate. He and Karl Popper authored a book defending dualism, called *The Self and Its Brain*, published in 1977. They rejected physical determinism in favor of free will: the mind could choose between different thoughts and courses of action undertaken by the brain and body.[13]

The most common objection to dualism is that it violates the conservation of energy: how can the mind, if it has no physical existence, initiate physical changes in the brain? Together with the German physicist Friedrich Beck, Eccles has provided the following answer: The brain's nerve cells fire when charged molecules, or ions, accumulate at a synapse, causing it to release neurotransmitters. But the presence of a given number of ions at a synapse does not always trigger the firing of a neuron. The reason, according to Eccles, is that for at least an instant, the ions exist in a quantum superposition of states; in some states the neuron discharges and in others it does not.

The mind exerts its influence over the brain by "deciding" which neurons will fire and which will not. As long as probability is conserved throughout the brain, this exercise of free will does not violate conservation of energy. "We have no proof of any of this," Eccles cheerfully acknowledged after explaining his theory to me in a telephone interview. He nonetheless called the hypothesis "a tremendous advance" that would inspire a resurgence of dualism. Materialism and all its vile progeny— logical positivism, behaviorism, identity theory (which equates states of mind with physical states of the brain)—"are finished," Eccles declared.

Eccles was frank—too frank for his own good—about his motive for turning to quantum mechanics in explaining the mind's properties. He was a "religious person," he told me, who rejected "cheap materialism." He believed that "the very nature of the mind is the same as the nature of life. It's a divine creation." Eccles also insisted that "we're only at the beginning of discovering the mystery of existence." Could we ever plumb that mystery, I asked, and thereby bring science to an end? "I don't think so," he replied. He paused and added, heatedly, "I don't *want* it to end. The only important thing is to go on." He agreed with his fellow dualist and falsificationist Karl Popper that we must and will keep "discovering and discovering and discovering. And thinking. And we must not claim to have the last word on anything."

What Roger Penrose Really Wants

Roger Penrose is somewhat better at obscuring his ulterior motives—perhaps because he perceives them only dimly himself. Penrose first made his reputation as an authority on black holes and other exotica of physics. He also invented Penrose tiles, simple geometric forms that when fit together generate infinitely varied, quasi-periodic patterns. Since 1989 he has been renowned for the arguments set forth in *The Emperor's New Mind*. The book's main purpose was to refute the claim of artificial-intelligence proponents that computers could replicate all the attributes of humans, including consciousness.

The key to Penrose's argument is Gödel's incompleteness theorem. The theorem states that any consistent system of axioms beyond a certain basic level of complexity yields statements that can be neither proved nor disproved with those axioms; hence the system is always incomplete. According to Penrose, the theorem implies that no "computable" model—that is, neither classical physics, computer science, nor neuroscience as presently construed—can replicate the mind's creative, or, rather, intuitive powers. The mind must derive its power from some more subtle phenomenon, probably one related to quantum mechanics.

Three years after my first meeting with Penrose in Syracuse—which had triggered my interest in the limits of science—I visited him at his home base, the University of Oxford. Penrose said he was working on a sequel to *Emperor's New Mind*, which would spell out his theory in greater detail. He was more convinced than ever, he said, that he was on the right track with his theory of quasi-quantum consciousness. "One is going out on a limb here, but I feel quite strongly about those things. I can't see any way out."[14]

I noted that some physicists had begun thinking about how exotic quantum effects, such as superposition, might be harnessed for performing computations that classical computers could not achieve. If these quantum computers proved feasible, would Penrose grant that they might be capable of thinking? Penrose shook his head. A computer capable of thought, he said, would have to rely on mechanisms related not to quantum mechanics in its present form but to a deeper theory not yet discovered. What he was *really* arguing against in *The Emperor's New Mind*, Penrose confided, was the assumption that the mystery of consciousness, or of reality in general, could be explained by the current laws of physics. "I'm saying this is wrong," he announced. "The laws that govern the behavior of the world are, I believe, much more subtle than that."

Contemporary physics simply does not make sense, he elaborated. Quantum mechanics, in particular, *has* to be flawed, because it is so glaringly inconsistent with ordinary, macroscopic reality. How can electrons act like particles in one experiment and waves in another? How can they be in two places at the same time? There must be some deeper theory that eliminates the paradoxes of quantum mechanics and its disconcertingly subjective elements. "Ultimately our theory has to accommodate subjectivism, but I wouldn't like to see the theory itself being a subjective theory." In other words, the theory should allow for the existence of minds but should not *require* it.

Neither superstring theory—which is after all a quantum theory—nor any other current candidate for a unified theory has the qualities that Penrose feels are necessary. "If there is going to be such a kind of total theory of physics, in some sense it couldn't conceivably be the character of any theory I've seen," he said. Such a theory would need a "kind of compelling naturalism." In other words, the theory would have to make sense.

Yet Penrose was just as conflicted as he had been in Syracuse over whether physics would achieve a truly complete theory. Gödel's theorem, he said, suggests that there will always be open questions in physics, as in mathematics. "Even if you could find the end of how the physical world actually was as a mathematical structure," Penrose remarked, "there would still be no end to the subject, if you like, just because there's no end to mathematics." He spoke very deliberately, much more so than when we first discussed the topic in 1989. He had clearly given the issue more thought.

I recalled Richard Feynman's comparison of physics to chess: once we learn the basic rules, we can still explore their consequences forever. "Yes, it's not dissimilar to that viewpoint," Penrose said. Did that mean he thought it would be possible to know the fundamental rules, if not all the consequences of those rules? "I suppose in my more optimistic moods I believe that." He added heatedly, "I'm certainly not one of these people who thinks there's no end to our physical understanding of the world." In Syracuse, Penrose had said it was pessimistic to believe in *The Answer*; now he felt this view was optimistic.

Penrose said he was pleased, for the most part, with the reception that had been given his ideas; most of his critics had been polite, at least. One exception was Marvin Minsky. Penrose had had an unpleasant encounter with Minsky at a meeting in Canada where they both lectured. At Minsky's insistence, Penrose spoke first. Minsky then rose to offer a rebuttal. After announcing that wearing a jacket "implies you're a gentleman," he took his jacket off, exclaiming, "Well, I don't feel like a gentleman!" He proceeded to attack *The Emperor's New Mind* with arguments that were, according to Penrose, silly. Recalling this scene, Penrose seemed still mystified, and pained. I marveled again, as I had when I first met him, at the contrast between Penrose's mildness of manner and the audacity of his intellectual views.

In 1994, two years after my meeting with Penrose in Oxford, his book *Shadows of the Mind* was published. In *The Emperor's New Mind*, Penrose had been rather vague about where quasi-quantum effects might work their magic. In *Shadows* he hazarded a guess: microtubules, minute tunnels of protein that serve as a kind of skeleton for most cells, including neurons. Penrose's hypothesis was based on a claim by Stuart Hameroff, an anesthesiologist at the University of Arizona, that anesthesia inhibits the movement of electrons in microtubules. Erecting a mighty theoretical edifice on this frail claim, Penrose conjectured that microtubules perform nondeterministic, quasi-quantum computations that somehow give rise to consciousness. Each neuron is thus not a simple switch, but a complex computer in its own right.

Penrose's microtubule theory could not help but be anticlimactic. In his first book, he built up an air of suspense, anticipation, and mystery, as does a director of a horror movie who offers only tantalizing glimpses of the monster. When Penrose finally unveiled his monster, it looked like an overweight actor wearing a cheap rubber suit, complete with flapping fins.

Some skeptics have responded, not unexpectedly, with ridicule rather than with awe. They have noted that microtubules are found in almost all cells, not just in neurons. Does that mean that our livers are conscious? What about our big toes? What about paramecia? (Penrose's partner, Stuart Hameroff, when I put this question to him in April 1994, replied, "I'm not going to contend that a paramecium is conscious, but it does show pretty intelligent behavior.")

Penrose can also be countered with Crick's argument against free will. Because Penrose, through mere introspection, cannot retrace the computational logic of his perception of a mathematical truth, he insists that the perceptions must stem from some mysterious, noncomputational phenomenon. But as Crick pointed out, just because we are not aware of the neural processes leading up to a decision does not mean that those processes did not occur. Proponents of artificial intelligence rebut Penrose's Gödel argument by contending that one can always design a computer to broaden its own base of axioms to solve a new problem; in fact, such learning algorithms are rather common (although they are still extremely crude in comparison to the human mind).[15]

Some critics of Penrose have accused him of being a vitalist, someone who secretly hopes that the mystery of the mind will not yield to science. But if Penrose was a vitalist, he would have kept his ideas vague and untestable. He would never have unveiled the microtubule monster. Penrose is a true scientist; he wants to *know*. He sincerely believes that our *present* understanding of reality is incomplete, logically flawed, and, well, mysterious. He is looking for a key, an insight, some clever quasi-quantum trick that will make everything suddenly clear. He is looking for *The Answer*. He has made the great mistake of thinking that physics should render the world completely intelligible and meaningful. Steven Weinberg could have told him that physics lacks that capacity.

Attack of the Mysterians

Penrose, although he pushed a theory of consciousness far beyond the horizon of current science, at least held out the hope that the theory could one day be reached. But some philosophers have questioned whether *any* purely materialistic model—involving conventional neural processes or the exotic, nondeterministic mechanisms envisioned by Penrose—can *really* account for consciousness. The philosopher Owen Flanagan named these doubters "the new mysterians," after the sixties rock group Question Mark and the Mysterians, which performed the hit song "96 Tears." (Flanagan himself is not a mysterian, but a down-to-earth materialist.)[16]

The philosopher Thomas Nagel offered one of the clearest expressions of the mysterian viewpoint (an oxymoronic undertaking?) in his famous 1974 essay "What Is It Like to Be a Bat?" Nagel assumed that subjective experience is a fundamental attribute of humans and many higher-level animals, such as bats. "No doubt it occurs in countless forms totally unimaginable to us, on other planets in other solar systems throughout the universe," Nagel wrote. "But no matter how the form may vary, the fact that an organism has conscious experience *at all* means, basically, that there is something it is like to *be* that organism."[17] Nagel argued that no matter how much we learn about the physiology of bats, we cannot *really* know what it is like to be one, because science cannot penetrate the realm of subjective experience.

Nagel is what one might call a weak mysterian: he holds out the possibility that philosophy and/or science might one day reveal a natural way to bridge the gap between our materialistic theories and subjective experience. Colin McGinn is a strong mysterian. McGinn is the same philosopher who believes that most major philosophical questions are unsolvable because they are beyond our cognitive abilities (see Chapter 2). Just as rats have cognitive limitations, so do humans, and one of our limitations is that we cannot solve the mind–body problem. McGinn considers his position on the mind–body problem—that it is unsolvable—the logical conclusion of Nagel's analysis in "What Is It Like to Be a Bat?" McGinn defends his viewpoint as superior to what he calls the "eliminativist" position, which attempts to show that the mind–body problem is not really a problem at all.

It is quite possible, McGinn said, for scientists to invent a theory of the mind that can predict the outcome of experiments with great precision and yield a wealth of medical benefits. But an effective theory is not necessarily a comprehensible one. "There's no real reason why part of our mind can't develop a formalism with these remarkable predictive properties, but we can't make sense of the formalism in terms of the part of our mind which understands things. So it could be in the case of consciousness that we come up with a theory which is analogous to quantum theory in this respect, a theory which is actually a good theory of consciousness, but we wouldn't be able to interpret it, or to understand it."[18]

This kind of talk infuriates Daniel Dennett. A philosopher at Tufts University, Dennett is a big, hoary-bearded man with a look of perpetual, twinkly amusement—Santa Claus on a diet. Dennett exemplifies what McGinn calls the eliminativist position. In his 1992 book, *Consciousness Explained,* Dennett contended that consciousness—and our sense that we possess a unified self—was an illusion arising out of the interaction of many different "subprograms" run on the brain's hardware.[19] Dennett, when I asked him about McGinn's mysterian argument, called it ridiculous. He denigrated McGinn's comparison of humans to rats. Unlike humans, Dennett asserted, rats cannot conceive of scientific questions, so of *course* they cannot solve them. Dennett suspected that McGinn and other mysterians "don't want consciousness to fall to science. They like the idea that this is off-limits to science. Nothing else could explain why they welcome such slipshod arguments."

Dennett tried another strategy, one that struck me as oddly Platonic for such an avowed materialist—and also dangerous for a writer. He recalled that Borges, in his story "The Library of Babel," imagined an infinite library of all possible statements, those that have been, will be, and could be, from the most nonsensical to the most sublime. Surely somewhere in the Library of Babel is a perfectly stated resolution of the mind–body problem, Dennett said. Dennett served up this argument with such confidence that I suspected he believed that the Library of Babel really exists.

Dennett granted that neuroscience might never produce a theory of consciousness that satisfied everyone. "We can't explain *anything* to everyone's satisfaction," he said. After all, many people are dissatisfied with science's explanations of, say, photosynthesis or biological reproduction. But "the sense of mystery is gone from photosynthesis or reproduction," Dennett said, "and I think in the end we will have a similar account of consciousness."

Abruptly, Dennett tacked in a completely different direction. "There's a curious paradox looming" in modern science, he said. "One of the very trends that makes science proceed so rapidly these days is a trend that leads science away from human understanding. When you switch from trying to model things with elegant equations to doing massive computer simulations . . . you may end up with a model that exquisitely models nature, the phenomena you're interested in, but you don't understand the model. That is, you don't understand it the way you understood models in the old days."

A computer program that accurately modeled the human brain, Dennett noted, might be as inscrutable to us as the brain itself. "Software systems are already at the very edge of human comprehensibility," he observed. "Even a system like the Internet is absolutely trivial compared to a brain, and yet it's been patched and built on so much that nobody really understands how it works or whether it will go on working. And thus, as you start using software-writing programs and software-debugging programs and code that heals itself, you create new artifacts that have a life of their own. And they become objects that are no longer within the epistemological hegemony of their makers. And so that's going to be sort of like the speed of light. It's going to be a barrier against which science is going to keep butting its head forever."

Astonishingly, Dennett was implying that he, too, had mysterian inclinations. He thought a theory of the mind, although it might be highly effective and have great predictive power, was unlikely to be intelligible to mere humans. The only hope humans have of comprehending their own complexity may be to cease being human. "Anybody who has the motivation or talent," he said, "will be able in effect to merge with these big software systems." Dennett was referring to the possibility, advanced by some artificial-intelligence enthusiasts, that one day we humans will be able to abandon our mortal, fleshy selves and become machines. "I think that's logically possible," Dennett added. "I'm not sure how plausible it is. It's a coherent future. I think it's not self-contradictory." But Dennett seemed to doubt whether even superintelligent machines would ever fully comprehend themselves. Trying to know themselves, the machines would have to become still more complicated; they would thus be caught in a spiral of ever-increasing complexity, chasing their own tails for all eternity.

How Do I Know You're Conscious?

In the spring of 1994, I witnessed a remarkable collision of the philosophical and scientific worldviews at a meeting called "Toward a Scientific Basis for Consciousness," held at the University of Arizona.[20] On the first day David Chalmers, a long-haired Australian philosopher who bears an uncanny resemblance to the subject of Thomas Gainsborough's famous painting *Blue Boy,* set forth the mysterian viewpoint in forceful terror. Studying neurons, he declared, cannot reveal why the impingement of sound waves on our ears gives rise to our subjective *experience* of Beethoven's Fifth Symphony. All physical theories, Chalmers said, describe only functions—such as memory, attention, intention, introspection—correlating to specific physical processes in the brain. But none of these theories can explain why the performance of these functions is accompanied by subjective experience. After all, one can certainly imagine a world of androids that resemble humans in every respect—except that they do not have a conscious experience of the world. However much they learn about the brain, neuroscientists cannot bridge that "explanatory gap" between the physical and subjective realms with a strictly physical theory, according to Chalmers.

Up to this point, Chalmers had expressed the basic mysterian viewpoint, the same one associated with Thomas Nagel and Colin McGinn. But then Chalmers proclaimed that although science could not solve the mind-body problem, philosophy still might. Chalmers thought he had found a possible solution: scientists should assume that information is as essential a property of reality as matter and energy. Chalmers's theory was similar to the it from bit concept of John Wheeler—in fact, Chalmers acknowledged his debt to Wheeler—and it suffered from the same fatal flaw. The concept of information does not make sense unless there is an information processor—whether an amoeba or a particle physicist—that gathers information and acts on it. Matter and energy were present at the dawn of creation, but life was not, as far as we know. How, then, can information be as fundamental as matter and energy? Nevertheless, Chalmers's ideas struck a chord among his audience. They thronged around him after his speech, telling him how much they had enjoyed his message.[21]

At least one listener was displeased: Christof Koch, Francis Crick's collaborator. That night Koch, a tall, rangy man wearing red cowboy boots, tracked Chalmers down at a cocktail party for the conferees and chastized him for his speech. It is precisely because philosophical approaches to consciousness have all failed that scientists must focus on the brain, Koch declared in his rapid-fire, German-accented voice as rubberneckers gathered. Chalmers's information-based theory of consciousness, Koch continued, like all philosophical ideas, was untestable and therefore useless. "Why don't you just say that when you have a brain the Holy Ghost comes down and makes you conscious!" Koch exclaimed. Such a theory was unnecessarily complicated, Chalmers responded drily, and it would not accord with his own subjective experience. "But how do I know your subjective experience is the same as mine?" Koch sputtered. "How do I even know you're conscious?"

Koch had brought up the embarrassing problem of solipsism, which lies at the heart of the mysterian position. No person really *knows* that any other being, human or inhuman, has a subjective experience of the world. By raising this ancient philosophical conundrum, Koch, like Dennett, was revealing himself to be a mysterian. Koch admitted as much to me later. All science can do, he asserted, is provide a detailed map of the physical processes that correlate with different subjective states. But science cannot truly "solve" the mind–body problem. No empirical, neurological theory can explain why mental functions are accompanied by specific subjective states. "I don't see how any science can explain that," Koch said. For the same reason, Koch doubted that science would ever provide a definitive answer to the question of whether machines can become conscious and have subjective experiences. "The debate might not be resolved ever," he told me, adding gratuitously, "How do I even know you are conscious?"

Even Francis Crick, although he was more optimistic than Koch, had to acknowledge that the solution to consciousness might not be intuitively comprehensible. "I don't think it will be a commonsense answer that we get when we understand the brain," Crick said. After all, natural selection cobbles organisms together not according to any logical plan, but with various gimmicks and tricks, with whatever works. Crick went on to suggest that the mysteries of the mind might not yield as readily as those of heredity. The mind "is a much more complicated system" than the genome, he remarked, and theories of the mind would probably have more limited explanatory power.

Holding up his pen, Crick explained that scientists should be able to determine which neural activity correlated with my perception of the pen. "But if you were to ask, 'Do you see red and blue the same way I see red and blue?' well, that's something you can't communicate to me. So I don't think we'll be able to explain everything that we're conscious of."

Just because the mind stems from deterministic processes, Crick continued, does not mean that scientists will be able to predict all its meanderings; they may be chaotic and thus unpredictable. "There may be other limitations in the brain. Who knows? I don't think you can look too far ahead." Crick doubted that quantum phenomena played a crucial role in consciousness, as Roger Penrose had suggested. On the other hand, Crick added, some neural equivalent of Heisenberg's uncertainty principle might restrict our ability to trace the brain's activity in minute detail, and the processes underlying consciousness might be as paradoxical and difficult for us to grasp as quantum mechanics is. "Remember," Crick elaborated, "our brains have evolved to deal with everyday matters when we were hunter-gatherers, and before that when we were monkeys." Yes, that was the point of Colin McGinn, and Chomsky, and Stent.

The Many Minds of Marvin Minsky

The most unlikely mysterian of all is Marvin Minsky. Minsky was one of the founders of artificial intelligence (AI), which holds that the brain is nothing more than a very complicated machine whose properties can be duplicated with computers. Before I visited him at MIT, colleagues warned me that he might be cranky, even hostile. If I did not want the interview cut short, I should not ask him too directly about the falling fortunes of artificial intelligence or of his own particular theories of the mind. One former associate pleaded with me not to take advantage of Minsky's penchant for outrageous utterances. "Ask him if he means it, and if he doesn't say it three times you shouldn't use it," the ex-colleague urged.

When I met Minsky, he was rather edgy, but the condition seemed congenital rather than acquired. He fidgeted ceaselessly, blinking, waggling his foot, pushing things about his desk. Unlike most scientific celebrities, he gave the impression of conceiving ideas and tropes from scratch rather than retrieving them whole from memory. He was often, but not always, incisive. "I'm rambling here," he muttered after a riff on how to verify models of the mind collapsed in a heap of sentence fragments.[22]

Even his physical appearance had an improvisational air. His large, round head seemed entirely bald, but was actually fringed by hairs as transparent as optical fibers. He wore a braided belt that supported, in addition to his pants, a belly pack and a tiny holster containing pliers with retractable jaws. With his paunch and vaguely Asian features, he resembled Buddha—Buddha reincarnated as a hyperactive hacker.

Minsky seemed unable, or unwilling, to inhabit any emotion for long.

Early on, as predicted, he lived up to his reputation as a curmudgeon, and arch-reductionist. He expressed contempt for those who doubted whether computers could be conscious. Consciousness is a trivial issue, he said. "I've solved it, and I don't understand why people don't listen." Consciousness is merely a type of short-term memory, a "low-grade system for keeping records." Computer programs such as LISP, which have features that allow their processing steps to be retraced, are "extremely conscious," more so than humans, with their pitifully shallow memory banks.

Minsky called Roger Penrose a "coward" who could not accept his own physicality, and he derided Gerald Edelman's reentrant-loops hypothesis as warmed-over feedback theory. Minsky even snubbed MIT's own Artificial Intelligence Laboratory, which he had founded and where we happened to be meeting. "I don't consider this to be a serious research institution at the moment," he announced.

When we wandered through the lab looking for a lecture on a chess-playing computer, however, a metamorphosis occurred. "Isn't the chess meeting supposed to be here?" Minsky queried a group of researchers chatting in a lounge. "That was yesterday," someone replied. After asking a few questions about the talk, Minsky spun tales about the history of chess-playing programs. This minilecture evolved into a reminiscence of Minsky's friend Isaac Asimov, who had just died. Minsky recounted how Asimov—who had popularized the term *robot* and explored its metaphysical implications in his science fiction—always refused Minsky's invitations to see the robots being built at MIT out of fear that his imagination "would be weighed down by this boring realism."

One lounger, noticing that he and Minsky wore the same pliers, yanked his instrument from its holster and with a flick of his wrist snapped the retractable jaws into place. "En garde," he said. Minsky, grinning, drew his weapon, and he and his challenger whipped their pliers repeatedly at each other, like punks practicing their switchblade technique. Minsky expounded on both the versatility and—an important point for him—the limitations of the pliers; his pair pinched him during certain maneuvers. "Can you take it apart with itself?" someone asked. Minsky and his colleagues shared an insiders' laugh at this reference to a fundamental problem in robotics.

Later, returning to Minsky's office, we encountered a young, extremely pregnant Korean woman. She was a doctoral candidate and was scheduled for an oral exam the next day. "Are you nervous?" asked Minsky. "A little," she replied. "You shouldn't be," he said, and gently pressed his forehead against hers, as if seeking to infuse her with his strength. I realized, watching this scene, that there were many Minskys.

But of course there would be. Multiplicity is central to Minsky's view of the mind. In his book *The Society of Mind* he contended that brains contain many different, highly specialized structures that evolved in order to solve different problems.[23] "We have many layers of networks of learning machines," he explained to me, "each of which has evolved to correct bugs or to adapt the other agencies to the problems of thinking." It is thus unlikely that the brain can be reduced to a particular set of principles or axioms, "because we're dealing with a real world instead of a mathematical one that is defined by axioms."

If AI had not lived up to its early promise, Minsky said, that was because modern researchers had succumbed to "physics envy"—the desire to reduce the intricacies of the brain to simple formulas. Such researchers, to Minsky's annoyance, had failed to heed his message that the mind has many different methods for coping with even a single, relatively simple problem. For example, someone whose television set fails to work will probably first consider the problem to be purely physical. He or she will check to see whether the television is properly programmed or whether the cord is plugged in. If that fails, the person may try to have the machine repaired, thus turning the problem from a physical one to a social one—how to find someone who can repair the television quickly and cheaply.

"That's one lesson I can't get across to these people," Minsky said of his fellow AI workers. "It seems to me that the problem the brain has more or less solved is how to organize different methods into working when the individual methods fail pretty often." The only theorist other than he who truly grasped the mind's complexity, Minsky asserted, was dead. "Freud has the best theories so far, next to mine, of what it takes to make a mind."

As Minsky continued speaking, his emphasis on multiplicity took on a metaphysical and even moral cast. He blamed the problems of his field—and of science in general—on what he called "the investment principle," which he defined as the tendency of humans to keep doing something that they have learned to do well rather than to move on to new problems. Repetition, or, rather, single-mindedness, seemed to hold a kind of horror for Minsky. "If there's something you like very much," he asserted, "then you should regard this not as you feeling good but as a kind of brain cancer, because it means that some small part of your mind has figured out how to turn off all the other things."

The reason Minsky had mastered so many skills during his career—he is an adept in mathematics, philosophy, physics, neuroscience, robotics, and computer science and has written several science fiction novels—was that he had learned to enjoy the "feeling of awkwardness" triggered by having to learn something new. "It's so thrilling not to be able to do something. It's such a rare experience to treasure. It won't last."

Minsky was a child prodigy in music, too, but eventually he decided that music was a soporific. "I think the reason people like music is to *suppress* thought—the wrong kinds of thought—not to produce it." Minsky still occasionally found himself composing "Bach-like things"—an electric piano crowded his office—but he tried to resist the impulse. "I had to kill the musician at some point," he said. "It comes back every now and then, and I hit it."

Minsky had no patience for those who claim the mind is too subtle to understand. "Look, before Pasteur people said, 'Life is different. You can't explain it mechanically.' It's just the same thing." But a final theory of the mind, Minsky emphasized, would be much more complex than a final theory of physics—which Minsky also believed was attainable. All of particle physics might be condensed to a page of equations, Minsky said, but to describe all the components of the mind would require much more space. After all, consider how long it would take precisely to describe an automobile, or even a single spark plug. "It would take a fair-sized book to explain how they welded and sintered the spline to the ceramic without it leaking when it starts."

Minsky said the truth of a model of mind could be demonstrated in several ways. First, a machine based on the model's principles should be able to mimic human development. "The machine ought to be able to start as a baby and grow up by seeing movies and playing with things." Moreover, as imaging technology improves, scientists should be able to determine whether the neural processes in living humans corroborate the model. "It seems to me that it's perfectly reasonable that once you get a [brain] scanner that had one angstrom [one ten-billionth of a meter] resolution, then you could see every neuron in someone's brain. You watch this for 1,000 years and you say, 'Well, we know exactly what happens whenever this person says "blue." ' And peole check this out for generations and the theory is sound. Nothing goes wrong, and that's the end of it."

If humans achieve a final theory of the mind, I asked, what frontiers will be left for science to explore? "Why are you asking me this question?" Minsky retorted. The concern that scientists will run out of things to do is pitiful, he said. "There's *plenty* to do." We humans may well be approaching our limits as scientists, but we will someday create machines much smarter than we that can continue doing science. But that would be machine science, not human science, I said. "You're a racist, in other words," Minsky said, his great domed forehead purpling. I scanned his face for signs of irony, but found none. "I think the important thing for us is to grow," Minsky continued, "not to remain in our own present stupid state." We humans, he added, are just "dressed up chimpanzees." Our task is not to preserve present conditions but to evolve, to create beings better, more intelligent than we.

But Minsky, surprisingly, was hard-pressed to say precisely what kinds of questions these brilliant machines might be interested in. Echoing Daniel Dennett, Minsky suggested, rather halfheartedly, that machines might try to comprehend themselves as they evolved into ever-more-complex entities. He seemed more enthusiastic discussing the possibilities of converting human personalities into computer programs that could then be downloaded into machines. Minsky saw downloading as a way to indulge in pursuits that he would ordinarily consider too dangerous, such as taking LSD or indulging in religious faith. "I regard religious experience as a very risky thing to do because it can destroy the brain in a rapid way, but if I had a backup copy—."

Minsky confessed that he would love to know what Yo-Yo Ma, the great cellist, felt like when playing a concerto, but Minsky doubted whether such an experience would be possible. To share Yo-Yo Ma's experience, Minsky explained, he would have to possess all Yo-Yo Ma's memories, he would have to *become* Yo-Yo Ma. But in becoming Yo-Yo Ma, Minsky suspected, he would cease to be Minsky.

This was an extraordinary admission for Minsky to make. Like literary critics who claim that the only true interpretation of a text is the text itself, Minsky was implying that our humanness is irreducible; any attempt to convert an individual into an abstract mathematical program—a string of ones and zeros that could be downloaded onto a disk and transferred from one machine to another or combined with another program representing another person—might well destroy that individual's essence. In his own oblique way, Minsky was suggesting that the how-do-I-know-you're-conscious problem is insurmountable. If no two personalities could ever be fully fused, then downloading, too, might be impossible. In fact, the whole premise of artificial intelligence, if intelligence is defined in human terms, might be flawed.

Minsky, for all his reputation as a rabid reductionist, is actually an *anti*reductionist. He is even more of a romantic, in his way, than is Roger Penrose. Penrose holds out the hope that the mind can be reduced to a single quasi-quantum trick. Minsky insists that no such reduction is possible, because multiplicity is the essence of the mind, of all minds, those of humans and machines alike. Minsky's revulsion toward single-mindedness, simplicity, reflects not just a scientific judgment, I think, but something deeper. Minsky, like Paul Feyerabend and David Bohm and other great romantics, seems to fear *The Answer*, the revelation to end all revelations. Fortunately for Minsky, no such revelation is likely to emerge from neuroscience, since any useful theory of the mind will probably be hideously complex, as he recognizes. Unfortunately for Minsky, it also seems unlikely, given this complexity, that he or even his grandchildren will witness the birth of machines with human attributes. If and when we do construct intelligent, autonomous machines, they will surely be aliens, as unlike us as a 747 is unlike a sparrow. And we could never be *sure* that they were conscious, any more than any of us knows that anyone else is conscious.

Did Bacon Solve the Consciousness Problem?

The conquest of consciousness will take time. The brain is marvelously complicated. But is it infinitely complicated? Given the rate at which neuroscientists are learning about it, within a few decades they may have a highly effective map of the brain, one that correlates specific neural processes to specific mental functions—including consciousness as defined by Crick and Koch. This knowledge may yield many practical benefits, such as treatments for mental illness and information-processing tricks that can be transferred to computers. In *The Coming of the Golden Age*, Gunther Stent proposed that advances in neuroscience might one day give us great power over our own selves. We might be able to "direct specific electrical inputs into the brain. These inputs can then be made to generate synthetically sensations, feelings, and emotions. . . . Mortal men will soon live like gods without sorrow of heart and remote from grief, as long as their pleasure centers are properly wired."[24]

But Stent, anticipating the mysterian arguments of Nagel, McGinn, and others, also wrote, "the brain may not be capable, in the last analysis, of providing an explanation of itself."[25] Scientists and philosophers will still strive to accomplish the impossible. They will ensure that neuroscience continues in a postempirical, ironic mode in which practitioners argue about the meaning of their physical models, much as physicists argue about the meaning of quantum mechanics. Every now and then a particularly evocative interpretation, set forth by some latter-day Freud steeped in neural and cybernetic knowledge, may attract a vast following and threaten to become the final theory of mind. Neomysterians will then sally forth and point out the theory's inevitable shortcomings. Can it provide a truly satisfying explanation of dreams, or mystical experience? Can it tell us whether amoebas are conscious, or computers?

One could argue that consciousness was "solved" as soon as someone decided that it was an epiphenomenon of the material world. Crick's blunt materialism echoes that of the British philosopher Gilbert Ryle, who coined the phrase "ghost in the machine" in the 1930s to ridicule dualism.[26] Ryle pointed out that dualism—which held that the mind was a separate phenomenon, independent of its physical substrate and capable of exerting influence over it—violated conservation of energy and thus all of physics. Mind is a property of matter, according to Ryle, and only by tracing the intricate meanderings of matter in the brain can one "explain" consciousness.

Ryle was not the first to propose this materialistic paradigm, which is at once so empowering and deflating. Four centuries ago, Francis Bacon urged the philosophers of his day to cease trying to show how the universe evolved from thought and to begin considering how thought evolved from the universe.[27] Here, arguably, Bacon anticipated modern explanations of consciousness within the context of the theory of evolution and, more generally, of the materialist paradigm. The scientific conquest of consciousness will be the ultimate anticlimax, yet another demonstration of Niels Bohr's dictum that science's job is to reduce all mysteries to trivialities. But human science will not, cannot, solve the how-do-I-know-you're-conscious problem. There may be only one way to solve it: to make all minds one mind.

CHAPTER EIGHT

The End of Chaoplexity

I miss the Reagan era. Ronald Reagan made moral and political choices so easy. What he liked, I disliked. Star Wars, for example. Formally known as the Strategic Defense Initiative, it was Reagan's plan to build a shield in space that would protect the United States from the nuclear missiles of the Soviet Union. Of the many stories I wrote about Star Wars, the one I am most embarrassed by now involved Gottfried Mayer-Kress, a physicist at, of all places, the Los Alamos National Laboratory, the cradle of the atomic bomb. Mayer-Kress had constructed a simulation of the arms race between the Soviet Union and the United States that employed "chaotic" mathematics. His simulation suggested that Star Wars would destabilize relations between the superpowers and possibly lead to a catastrophe, that is, nuclear war. Because I approved of Mayer-Kress's conclusions—and because his place of employment added a nice touch of irony—I wrote up an admiring report of his work. Of course, if Mayer-Kress's simulation had suggested that Star Wars was a good idea, I would have dismissed his work as the nonsense that it obviously was. Star Wars could well have destabilized relations between the superpowers, but did we need some computer model to tell us that?

I don't mean to beat up on Mayer-Kress. He meant well. (In 1993, several years after I wrote about Mayer-Kress's Star Wars research, I saw a press release from the University of Illinois, where he was then employed, announcing that his computer simulations had suggested solutions to the conflicts in Bosnia and Somalia.)[1] His work is just one of the more blatant examples of over-reaching by someone in the field of chaoplexity (pronounced kay-oh-*plex*-ity). By chaoplexity I mean both chaos and its close relative complexity. Each term, and chaos in particular, has been defined in specific, distinct ways by specific individuals. But each has also been defined in so many overlapping ways by so many different scientists and journalists that the terms have become virtually synonymous, if not meaningless.

The field of chaoplexity emerged as a full-blown pop-culture phenomenon with the publication in 1987 of *Chaos: Making a New Science*, by former *New York Times* reporter James Gleick. After Gleick's masterful book became a best-seller, scores of journalists and scientists sought to duplicate his success by writing similar books on similar topics.[2] There are two, somewhat contradictory aspects to the chaoplexity message. One is that many phenomena are nonlinear and hence inherently unpredictable, because arbitrarily tiny influences can have enormous, unforeseeable consequences. Edward Lorenz, a meteorologist at MIT and a pioneer of chaoplexity, called this phenomenon the butterfly effect, because it meant that a butterfly fluttering in Iowa could, in principle, trigger an avalanche of effects culminating in a monsoon in Indonesia. Because we can never possess more than approximate knowledge of a weather system, our ability to predict its behavior is severely limited.

This insight is hardly new. Henri Poincaré warned at the turn of the century that "small differences in the initial conditions produce very great ones in the final phenomena. A small error in the former will produce an enormous error in the latter. Prediction becomes impossible."[3] Investigators of chaoplexity—whom I will call chaoplexologists—also like to emphasize that many phenomena in nature are "emergent"; they exhibit properties that cannot be predicted or understood simply by examining the system's parts. Emergence, too, is a hoary idea, related to holism, vitalism, and other antireductionist creeds that date back to the last century at least. Certainly Darwin did not think that natural selection could be derived from Newtonian mechanics.

So much for the negative side of the chaoplexity message. The positive side goes as follows: The advent of computers and of sophisticated nonlinear mathematical techniques will help modern scientists understand chaotic, complex, emergent phenomena that have resisted analysis by the reductionist methods of the past. The blurb on the back of Heinz Pagels's *The Dreams of Reason*, one of the best books on the "new sciences of complexity," put it this way: "Just as the telescope opened up the universe and the microscope revealed the secrets of the microcosm, the computer is now opening an exciting new window on the nature of reality. Through its capacity to process what is too complex for the unaided mind, the computer enables us for the first time to simulate reality, to create models of complex systems like large molecules, chaotic systems, neural nets, the human body and brain, and patterns of evolution and population growth."[4]

This hope stems in large part from the observation that simple sets of mathematical instructions, when carried out by a computer, can yield fantastically complicated and yet strangely ordered effects. John von Neumann may have been the first scientist to recognize this capability of computers. In the 1950s, he invented the cellular automaton, which in its simplest form is a screen divided into a grid of cells, or squares. A set of rules relates the color, or state, of each cell to the state of its immediate neighbors. A change in the state of a single cell can trigger a cascade of changes throughout the entire system. "Life," created in the early 1970s by the British mathematician John Conway, remains one of the most celebrated of cellular automatons. Whereas most cellular automatons eventually settle into predictable, periodic behavior, Life generates an infinite variety of patterns—including cartoonlike objects that seem to be engaged in inscrutable missions. Inspired by Conway's strange computer world, a number of scientists began using cellular automatons to model various physical and biological processes.

Another product of computer science that seized the imagination of the scientific community was the Mandelbrot set. The set is named after Benoit Mandelbrot, an applied mathematician at IBM who is one of the protagonists of Gleick's book *Chaos* (and whose work on indeterministic phenomena led Gunther Stent to conclude that the social sciences would never amount to much). Mandelbrot invented fractals, mathematical objects displaying what is known as fractional dimensionality: they are fuzzier than a line but never quite fill a plane. Fractals also display patterns that keep recurring at finer and finer scales. After coining the term fractal, Mandelbrot pointed out that many real-world phenomena—notably clouds, snowflakes, coastlines, stock-market fluctuations, and trees—have fractal-like properties.

The Mandelbrot set, too, is a fractal. The set corresponds to a simple mathematical function that is repeatedly iterated; one solves the function and plugs the answer back into it and solves it again, ad infinitum. When plotted by a computer, the numbers generated by the function cluster into a now-famous shape, which has been likened to a tumor-ridden heart, a badly burned chicken, and a warty figure eight lying on its side. When one magnifies the set with a computer, one finds that its borders do not form crisp lines, but shimmer like flames. Repeated magnification of the borders plunges the viewer into a bottomless phantasmagoria of baroque imagery. Certain patterns, such as the basic heartlike shape, keep recurring, but always with subtle variations.

The Mandelbrot set, which has been called "the most complex object in mathematics," has become a kind of laboratory in which mathematicians can test ideas about the behavior of nonlinear (or chaotic, or complex) systems. But what relevance do those findings have to the real world? In his 1977 magnum opus, *The Fractal Geometry of Nature*, Mandelbrot warned that it was one thing to observe a fractal pattern in nature and quite another to determine the *cause* of that pattern. Although exploring the consequences of self-similarity yielded "extraordinary surprises, helping me to understand the fabric of nature," Mandelbrot said, his attempts to unravel the causes of self-similarity "had few charms."[5]

Mandelbrot seemed to be alluding to the seductive syllogism that underlies chaoplexity. The syllogism is this: There are simple sets of mathematical rules that when followed by a computer give rise to extremely complicated patterns, patterns that never quite repeat themselves. The natural world also contains many extremely complicated patterns that never quite repeat themselves. Conclusion: Simple rules underly many extremely complicated phenomena in the world. With the help of powerful computers, chaoplexologists can root out those rules.

Of course, simple rules *do* underlie nature, rules embodied in quantum mechanics, general relativity, natural selection, and Mendelian genetics. But chaoplexologists insist that much more powerful rules remain to be found.

The 31 Flavors of Complexity

Blue and red dots skittered across a computer screen. But these were not just colored dots. These were agents, simulated people, doing the things that real people do: foraging for food, seeking mates, competing and cooperating with each other. At least, that's what Joshua Epstein, the creator of this computer simulation, claimed. Epstein, a sociologist from the Brookings Institution, was showing his simulation to me and two other journalists at the Santa Fe Institute, where Epstein was a visiting fellow. The institute, founded in the mid-1980s, quickly became the headquarters of complexity, the self-proclaimed successor to chaos as the new science that would transcend the stodgy reductionism of Newton, Darwin, and Einstein.

As my colleagues and I watched Epstein's colored dots and listened to his even more colorful interpretation of their movements, we offered polite murmurs of interest. But behind his back we exchanged jaded smiles. None of us took this kind of thing very seriously. We all understood, implicitly, that this was ironic science. Epstein himself, when pressed, acknowledged that his model was not predictive in any way; he called it a "laboratory," a "tool," a "neural prosthesis" for exploring ideas about the evolution of societies. (These were all favorite terms of Santa Fe'ers.) But during public presentations of his work, Epstein had also claimed that simulations such as his would revolutionize the social sciences, helping to solve their most intractable problems.[6]

Another believer in the power of computers is John Holland, a computer scientist with joint appointments at the University of Michigan and the Santa Fe Institute. Holland was the inventor of genetic algorithms, which are segments of computer code that can rearrange themselves to produce a new program that can solve a problem more efficiently. According to Holland, the algorithms are, in effect, evolving, just as the genes of living organisms evolve in response to the pressure of natural selection.

Holland has proposed that it may be possible to construct a "unified theory of complex adaptive systems" based on mathematical techniques such as those embodied in his genetic algorithms. He spelled out his vision in a 1993 lecture:

Many of our most troubling long-range problems—trade imbalances, sustainability, AIDS, genetic defects, mental health, computer viruses—center on certain systems of extraordinary complexity. The systems that host these problems—economies, ecologies, immune systems, embryos, nervous systems, computer networks—appear to be as diverse as the problems. Despite appearances, however, the systems do share significant characteristics, so much so that we group them under a single classification at the Santa Fe Institute, calling them *complex adaptive systems* (cas). This is more than terminology. It signals our intuition that there are general principles that govern all cas behavior, principles that point to ways of solving the attendant problems. Much of our work is aimed at turning this intuition into fact.[7]

The ambition revealed by this statement is breathtaking. Chaoplexologists often ridicule particle physicists for their hubris, for thinking they can achieve a theory of everything. But actually particle physicists are rather modest in their ambition: they merely hope that they can wrap up the forces of nature in one tidy package and perhaps illuminate the origin of the universe. Few are so bold as to propose that their unified theory will yield both truth (that is, insight into nature) and *happiness* (solutions to our worldly problems), as Holland and others have proposed. And Holland is considered one of the more modest scientists associated with the field of complexity.

But can scientists achieve a unified theory of complexity if they cannot agree what, precisely, complexity is? Students of complexity have struggled, with little success, to distinguish themselves from those who study chaos. According to University of Maryland physicist James Yorke, chaos refers to a restricted set of phenomena that evolve in predictably unpredictable ways—demonstrating sensitivity to initial conditions, aperiodic behavior, the recurrence of certain patterns at different spatial and temporal scales, and so on. (Yorke ought to know, since he coined the term *chaos* for a paper published in 1975.) Complexity seems to refer to "anything you want," according to Yorke.[8]

One widely touted definition of complexity involves "the edge of chaos." This picturesque phrase was incorporated in the subtitles of two books by journalists published in 1992: *Complexity: Life at the Edge of Chaos*, by Roger Lewin, and *Complexity: The Emerging Science at the Edge of Order and Chaos*, by M. Mitchell Waldrop.[9] (The authors no doubt intended the phrase to evoke the style as well as the substance of the field.) The basic idea of the edge of chaos is that nothing novel can emerge from systems with high degrees of order and stability, such as crystals; on the other hand, completely chaotic, or aperiodic, systems, such as turbulent fluids or heated gases, are *too* formless. Truly complex things—amoebas, bond traders, and the like—happen at the border between rigid order and randomness.

Most popular accounts credit this idea to the Santa Fe researchers Norman Packard and Christopher Langton. Packard, whose experience as a leading figure in chaos theory educated him in the importance of idea packaging, coined the all-important phrase "edge of chaos" in the late 1980s. In experiments with cellular automatons, he and Langton concluded that a system's computational potential—that is, its ability to store and process information—peaks in a regime between highly periodic and chaotic behavior. But two other researchers at the Santa Fe Institute, Melanie Mitchell and James Crutchfield, have reported that their own computer experiments do not support the conclusions of Packard and Langton. They also questioned whether "anything like a drive toward universal-computational capabilities is an important force in the evolution of biological organisms."[10] Although a few Santa Fe'ers still employ the phrase "edge of chaos" (notably Stuart Kauffman, whose work was described in Chapter 5), most now disavow it.

Many other definitions of complexity have been proposed—at least 31, according to a list compiled in the early 1990s by physicist Seth Lloyd of the Massachusetts Institute of Technology, who is also associated with the Santa Fe Institute.[11] The definitions typically draw on thermodynamics, information theory, and computer science and involve concepts such as entropy, randomness, and information—which themselves have proved to be notoriously slippery terms. All definitions of complexity have drawbacks. For example, algorithmic information theory, proposed by the IBM mathematician Gregory Chaitin (among others), holds that the complexity of a system can be represented by the shortest computer program describing it. But according to this criterion, a text created by a team of typing monkeys is more complex—because it is more random and therefore less compressible—than *Finnegans Wake*.

Such problems highlight the awkward fact that complexity exists, in some murky sense, in the eye of a beholder (like pornography, for instance).[12] Researchers have at times debated whether complexity has become so meaningless that it should be abandoned, but they have invariably concluded that the term has too much public relations value. Santa Fe'ers often employ "interesting" as a synonym for "complex." But what government agency would supply funds for research on a "unified theory of interesting things"?

The Poetry of Artificial Life

Members of the Santa Fe Institute may not agree on what they are studying, but they concur on how they should study it: with computers. Christopher Langton embodies the faith in computers that gave rise to the chaos and complexity movements. He has proposed that simulations of life run on a computer are alive—not sort of, or in a sense, or metaphorically, but actually. Langton is the founding father of artificial life, a subfield of chaoplexity that has attracted much attention in its own right. Langton has helped organize several conferences on artificial life—the first held at Los Alamos in 1987—attended by biologists, computer scientists, and mathematicians who share his affinity for computer animation.[13]

Artificial life is an outgrowth of artificial intelligence, a field that preceded it by several decades. Whereas artificial-intelligence researchers seek to understand the mind better by simulating it on a computer, proponents of artificial life hope to gain insights into a broad range of biological phenomena through their simulations. And just as artificial intelligence has generated more portentous rhetoric than tangible results, so has artificial life. As Langton stated in an essay introducing the inaugural issue of the quarterly journal *Artificial Life* in 1994:

> Artificial life will teach us much about biology—much that we could not have learned by studying the natural products of biology alone—but artificial life will ultimately reach beyond biology, into a realm we do not yet have a name for, but which must include culture and our technology in an extended view of nature. I don't want to paint a rosy picture of the future of artificial life. It will not solve all our problems. Indeed, it may well add to them. . . . Perhaps the simplest way to emphasize this point is by merely pointing out that Mary Shelley's prophetic story of Dr. Frankenstein can no longer be considered science fiction.[14]

Even before I met Langton, I felt I knew him. He played a prominent role in several popular journalistic treatments of chaoplexity. And no wonder.

He is the archetypal hip young scientist: simultaneously intense and mellow, longhaired, given to wearing jeans, leather vests, hiking boots, Indian jewelry. He also comes equipped with a marvelous life story; its centerpiece is a hang-gliding accident that led to a coma and an epiphany. (Those who want to hear the tale can read Waldrop's *Complexity*, Lewin's *Complexity*, or Steven Levy's *Artificial Life*.)[15]

When I finally encountered Langton in the flesh in Santa Fe in May 1994, we decided to talk while eating lunch at one of his favorite restaurants. Langton's car—hadn't I read about it in one of the books about him?—was a beat-up old compact filled with miscellanea, from audiotapes and pliers to plastic containers of hot sauce, all covered with a layer of beige desert dust. As we drove to the restaurant, Langton dutifully banged the chaoplexity boilerplate. Most scientists from Newton on have studied systems exhibiting stability, periodicity, and equilibrium, but he and other researchers at Santa Fe wanted to understand the "transient regimes" underlying many biological phenomena. After all, he said, "once you reach the equilibrium point for a living organism, you're dead."

He grinned. It had begun raining, and he turned on his wipers. The windshield quickly went from translucent to opaque, as the wipers smeared dirt across the glass. Langton, peering through an unsmeared corner of the glass, continued talking, seemingly unperturbed by the windshield's metaphorical message. Science, he said, had obviously made enormous progress by breaking things up into pieces and studying those pieces. But that methodology provided only limited understanding of higher-level phenomena, which were created to a large extent through historical accidents. One could transcend those limitations, however, through a synthetic methodology, in which the basic components of existence were put together in new ways in computers to explore what might have happened or could have happened.

"You end up with a much larger set" of possibilities, Langton said. "You can then probe the set not just of existing chemical compounds but of possible chemical compounds. And it's only really within that ground of the possible chemical compounds that you're going to see any regularity. The regularity is there but you can't see it in the very small set of things that nature initially provided you with." With computers, biologists can explore the role of chance by simulating the beginning of life on earth, altering the conditions and observing the consequences. "So part of what artificial life is all about, and part of the broader scheme that I just call synthetic biology in general, is probing beyond, pushing beyond the envelope of what occurred naturally." In this way, Langton suggested, artificial life might reveal which aspects of our history were inevitable and which were merely contingent.

In the restaurant, as he chewed on chicken fajitas, Langton confirmed that yes, he really did adhere to the view known as "strong a-life," which holds that computer simulations of living things are themselves alive. He described himself as a functionalist, who believed life was characterized by what it did rather than by what it was made of. If a programmer created molecule-like structures that, following certain laws, spontaneously organized themselves into entities that could seemingly eat, reproduce, and evolve, Langton would consider those entities to be alive—"even if they're in a computer."

Langton said his belief had moral consequences. "I like to think that if I saw somebody sitting next to me at a computer terminal who is *torturing* these creatures, you know, sending them to some digital equivalent of hell, or rewarding only a select few who spelled out his name on the screen, I would try to get this guy some psychological help!"

I told Langton that he seemed to be conflating metaphor, or analogy, with reality. "What I'm trying to do, actually, is something a little more seditious than that," Langton replied, smiling. He wanted people to realize that life might be a process that could be implemented by any number of arrangements of matter, including the ebb and flow of electrons in a computer. "At some level the actual physical realization is irrelevant to the functional properties," he said. "Of course there are differences," he added. "There are going to be differences if there's a different material base. But are the differences fundamental to the property of being alive or not?"

Langton did not support the claim—commonly made by artificial-intelligence enthusiasts—that computer simulations can also have subjective experiences. "This is why I like artificial life more than AI," he said. Unlike most biological phenomena, subjective states cannot be reduced to mechanical functions. "No mechanical explanation you could possibly give is going to give you that explanation for this sense of awareness, of I-ness, of my being here now." Langton, in other words, was a mysterian, someone who believed that an explanation of consciousness was beyond the reach of science. He finally conceded that the question of whether computer simulations are *really* alive was also, ultimately, a philosophical, and therefore unresolvable, issue. "But for artificial life to do its job and help broaden the empirical database for biological science and for a theory of biology, they don't have to solve that problem. Biologists have never really had to solve that."

The longer Langton spoke, the more he seemed to acknowledge—and even welcome the fact—that artificial life would never be the basis for a truly empirical science. Artificial-life simulations, he said, "force me to look over my shoulder about the assumptions I make about the real world." In other words, the simulations can enhance our negative capability; they can serve to challenge rather than to support theories about reality. Moreover, scientists studying artificial life might have to settle for something less than the "complete understanding" they derived from the old, reductionist methods. "For certain categories in nature there won't be anything more we can do by way of an explanation than to be able to say, 'Well, here's the history.' "

Then he confessed that such an outcome would suit him fine; he hoped the universe was in some fundamental sense "irrational." "Rationality is very much connected with the tradition in science for the last 300 years, when you're going to end up with some sort of understandable explanation of something. And I would be disappointed if that were the case."

Langton complained that he was frustrated with the linearity of scientific language. "There's a reason for poetry," he said. "Poetry is a very nonlinear use of language, where the meaning is more than just the sum of the parts. And science requires that it be nothing more than the sum of the parts. And just the fact that there's stuff to explain out there that's more than the sum of the parts means that the traditional approach, just characterizing the parts and the relations, is not going to be adequate for capturing the essence of many systems that you would like to be able to do. That's not to say that there isn't a way to do it in a more scientific way than poetry, but I just have the feeling that culturally there's going to be more of something like poetry in the future of science."

The Limits of Simulation

In February 1994, the journal *Science* published a paper, "Verification, Validation, and Confirmation of Numerical Models in the Earth Sciences," that addressed the problems posed by computer simulations. The remarkably postmodern article was written by Naomi Oreskes, a historian and geophysicist at Dartmouth College; Kenneth Belitz, a geophysicist also at Dartmouth; and Kristin Shrader Frechette, a philosopher at the University of South Florida. Although they focused on geophysical modeling, their warnings were really applicable to numerical models of all kinds (as they acknowledged in a letter printed by *Science* some weeks later).[16]

The authors observed that numerical models were becoming increasingly influential in debates over global warming, the depletion of oil reserves, the suitability of nuclear-waste sites, and other issues. Their paper was meant to serve as a warning that "verification and validation of numerical models of natural systems is impossible." The only propositions that can be verified—that is, proved true—are those dealing in pure logic or mathematics. Such systems are closed, in that all their components are based on axioms that are true by definition. Two plus two equals four by common agreement, not because the equation corresponds to some external reality. Natural systems are always open, Oreskes and her colleagues pointed out; our knowledge of them is always incomplete, approximate, at best, and we can never be sure we are not overlooking some relevant factors.

"What we call data," they explained, "are inference-laden signifiers of natural phenomena to which we have incomplete access. Many inferences and assumptions can be justified on the basis of experience (and some uncertainties can be estimated), but the degree to which our assumptions hold in any new study can never be established a priori. The embedded assumptions thus render the system open." In other words, our models are always idealizations, approximations, guesses.

The authors emphasized that when a simulation accurately mimics or even predicts the behavior of a real phenomenon, the model is still not verified. One can never be sure whether a match stems from some genuine correspondence between the model and reality or is coincidental. Moreover, it is always possible that other models, based on different assumptions, could yield the same results.

Oreskes and her coauthers noted that the philosopher Nancy Cartwright called numerical models "a work of fiction." They continued,

While not necessarily accepting her viewpoint, we might ponder this aspect of it: A model, like a novel, may resonate with nature, but it is not a "real" thing. Like a novel, a model may be convincing—it may ring true if it is consistent with our experience of the natural world. But just as we may wonder how much the characters of a novel are drawn from real life and how much is artifice, we might ask the same of a model: How much is based on observation and measurement of accessible phenomena, how much is based on informed judgement, and how much is convenience? . . . [We] must admit that a model may confirm our biases and support incorrect intuitions. Therefore, models are most useful when they are used to challenge existing formulations, rather than to validate or verify them.

Numerical models work better in some cases than in others. They work particularly well in astronomy and particle physics, because the relevant objects and forces conform to their mathematical definitions so precisely. Moreover, mathematics helps physicists define what is otherwise undefinable. A quark is a purely mathematical construct. It has no meaning apart from its mathematical definition. The properties of quarks—charm, color, strangeness—are mathematical properties that have no analogue in the macroscopic world we inhabit. Mathematical theories are less compelling when applied to more concrete, complex phenomena, such as anything in the biological realm. As the evolutionary biologist Ernst Mayr has pointed out, each organism is unique; each one also changes from moment to moment.[17] That is why mathematical models of biological systems generally have less predictive power than physics does. We should be equally wary of their ability to yield truths about nature.

The Self-Organized Criticality of Per Bak

This kind of "wishy-washy" philosophical skepticism irks Per Bak. Bak, a Danish physicist who came to the United States in the 1970s, is like a parody of Harold Bloom's strong poet. He is a tall, stout man, at once owlish and pugnacious, who bristles with opinions. Trying to convince me of the superiority of complexity to other modes of science, he scoffed at the suggestion that particle physicists could uncover the secret of existence by probing ever-smaller scales of matter. "The secret doesn't come from going deeper and deeper into the system," Bak asserted with a distinct Danish accent. "It comes from going the *other* direction."[18]

Particle physics is dead, Bak proclaimed, killed by its own success. Most particle physicists, he noted, "think they're still doing science when they're really just cleaning up the mess after the party." The same was true of solid-state physics, the field in which Bak began his career. The fact that thousands of physicists were working on high-temperature superconductivity—for the most part in vain—showed how desiccated the field had become: "There's very little meat and many animals who want to eat it." As for chaos (which Bak defined in the same narrow way that James Yorke did), physicists had come to a basic understanding of the processes underlying chaotic behavior by 1985, two years before Gleick's book *Chaos* was published. "That's how things go!" Bak barked. "Once something reaches the masses, it's already done." (Complexity, of course, is the exception to Bak's rule.)

Bak had nothing but contempt for scientists who were content merely to refine and extend the work of the pioneers. "There's no need for that! We don't need the cleanup team here!" Fortunately, Bak said, many mysterious phenomena continue to resist scientific understanding: the evolution of species, human cognition, economics. "What these things have in common is that they are very large things with many degrees of freedom. They are what we call complex systems. And there will be a revolution in science. These things will be made into hard sciences in the next years in the same way that [particle] physics and solid-state physics were made hard sciences in the last 20 years." Bak rejected the "pseudophilosophical, pessimistic, wishy-washy" view that these problems are simply too difficult for our puny human brains. "If I thought that was true I wouldn't be doing these things!" Bak exclaimed. "We should be optimistic, concrete, and then we can go on. And I'm sure that science will look totally different 50 years from now than it does today."

In the late 1980s Bak and two colleagues proposed what quickly became a leading candidate for a unified theory of complexity: self-organized criticality. His paradigmatic system is a sandpile. As one adds sand to the top of the pile, it approaches what Bak calls the critical state, in which even a single additional grain of sand dropped on the top of the pile can trigger an avalanche down the pile's sides. If one plots the size and frequency of the avalanches occurring in this critical state, the results conform to what is known as a power law: the frequency of avalanches is inversely proportional to a power of their size.

Bak credited the chaos pioneer Benoit Mandelbrot with having pointed out that earthquakes, stock-market fluctuations, the extinction of species, and many other phenomena displayed the same pattern of power-law behavior. (In other words, the phenomena that Bak defined as complex were also all chaotic.) "Since economics, geophysics, astronomy, biology have these singular features there *must* be a theory here," Bak said. He hoped his theory might explain why small earthquakes are common and large ones uncommon, why species persist for millions of years and then vanish, why stock markets crash. "We can't explain everything about everything, but something about everything."

Bak thought models such as his would eventually revolutionize economics. "Traditional economics is not a real science. It's a mathematical discipline where they talk about perfect markets and perfect rationality and perfect equilibrium." This approach was a "grotesque approximation" that could not explain real-world economic behavior. "Any real person who works on Wall Street and observes what happens knows that fluctuations come from chain reactions in the system. It comes from the coupling between the various agents: bank traders, customers, thieves, robbers, governments, economies, whatever. Traditional economics has no description of this phenomenon."

Can mathematical theories provide meaningful insights into cultural phenomena? Bak groaned at the question. "I don't understand what meaning is," he said. "In science there is no meaning to anything. It just observes and describes. It doesn't ask the atom why it's going left when it's subjected to a magnetic field. So social scientists should go out and observe the behavior of people and then figure out what the consequences of that are for society."

Bak acknowledged that such theories offered statistical descriptions rather than specific predictions. "The whole idea is we cannot predict. But nevertheless we can understand those systems that we cannot predict. We can understand *why* they cannot be predicted." That, after all, was what thermodynamics and quantum mechanics, which are also probabilistic theories, had accomplished. The models should be specific enough to be falsifiable, but not too specific, Bak said. "I think it's a losing game to make very specific and detailed models. That doesn't give any insight." That would be mere engineering, Bak sniffed.

When I asked whether he thought researchers would someday converge on a single, true theory of complex systems, Bak seemed to lose a bit of his nerve. "It is a much more fluid situation," he said. He doubted whether scientists would ever achieve a simple, unique theory of the brain, for instance. But they might find "some principles, hopefully not too many, that govern the behavior of the brain." He mused for a moment, then added, "I think it is a much more long-term thing than, say, chaos theory."

Bak also feared that the federal government's growing antipathy toward pure science and its increased emphasis on practical applications might hinder progress in complexity studies. It was increasingly difficult to pursue science for its own sake; science had to be useful. Most scientists were forced to do "deadly boring stuff that cannot be of any real interest." His own primary employer, Brookhaven National Laboratory, was making people do "*horrifying* things, *incredible* garbage." Even Bak, as fiercely optimistic as he was, had to acknowledge the dismal plight of modern science.

Self-organized criticality has been touted by, among others, Al Gore. In his 1992 best-seller, *Earth in the Balance*, Gore revealed that self-organized criticality had helped him to understand not only the sensitivity of the environment to potential disruptions but also "change in my own life."[19] Stuart Kauffman has found affinities between self-organized criticality, the edge of chaos, and the laws of complexity he has glimpsed in his computer simulations of biological evolution. But other researchers have complained that Bak's model does not even provide a very good description of his paradigmatic system: a sandpile. Experiments by physicists at the University of Chicago have shown that sandpiles behave in many different ways, depending on the size and shape of the grains; few sandpiles display the power-law behavior predicted by Bak.[20] Moreover, Bak's model may be too general and statistical in nature really to illuminate any of the systems it describes. After all, many phenomena can be described by a so-called Gaussian curve, more commonly known as a bell curve. But few scientists would claim, for example, that human intelligence scores and the apparent luminosity of galaxies must derive from common mechanisms.

Self-organized criticality is not really a theory at all. Like punctuated equilibrium, self-organized criticality is merely a description, one of many, of the random fluctuations, the noise, permeating nature. By Bak's own admission, his model can generate neither specific predictions about nature nor meaningful insights. What good is it, then?

Cybernetics and Other Catastrophes

History abounds with failed attempts to create a mathematical theory that explains and predicts a broad range of phenomena, including social ones. In the seventeenth century Leibniz fantasized about a system of logic so compelling that it could resolve not only all mathematical questions but also philosophical, moral, and political ones.[21] Leibniz's dream has persisted even in the century of doubt. Since World War II scientists have become temporarily infatuated with at least three such theories: cybernetics, information theory, and catastrophe theory.

Cybernetics was created largely by one person, Norbert Wiener, a mathematician at the Massachusetts Institute of Technology. The subtitle of his 1948 book, *Cybernetics*, revealed his ambition: *Control and Communication in the Animal and the Machine.*[22] Wiener based his neologism on the Greek term *kubernetes*, or steersman. He proclaimed that it should be possible to create a single, overarching theory that could explain the operation not only of machines but also of all biological phenomena, from single-celled organisms up through the economies of nation-states. All these entities process and act on information; they all employ such mechanisms as positive and negative feedback and filters to distinguish signals from noise.

By the 1960s cybernetics had lost its luster. The eminent electrical engineer John R. Pierce noted drily in 1961 that "in this country the word cybernetics has been used most extensively in the press and in popular and semiliterary, if not semiliterate, magazines."[23] Cybernetics still has a following in isolated enclaves, notably Russia (which during the Soviet era was highly receptive to the fantasy of society as a machine that could be fine-tuned by following the precepts of cybernetics). Wiener's influence persists in U.S. pop culture if not within science itself: we owe words such as *cyberspace*, *cyberpunk*, and *cyborg* to Wiener.

Closely related to cybernetics is information theory, which Claude Shannon, a mathematician at Bell Laboratories, spawned in 1948 with a two-part paper titled "A Mathematical Theory of Communication."[24] Shannon's great achievement was to invent a mathematical definition of information based on the concept of entropy in thermodynamics. Unlike cybernetics, information theory continues to thrive—within the niche for which it was intended. Shannon's theory was designed to improve the transmission of information over a telephone or telegraph line subject to electrical interference, or noise. The theory still serves as the theoretical foundation for coding, compression, encryption, and other aspects of information processing.

By the 1960s information theory had infected other disciplines outside communications, including linguistics, psychology, economics, biology, and even the arts. (For example, various sages tried to concoct formulas relating the quality of music to its information content.) Although information theory is enjoying a renaissance in physics as a result of the influence of John Wheeler (the it from bit) and others, it has yet to contribute to physics in any concrete way. Shannon himself doubted whether certain applications of his theory would come to much. "Somehow people think it can tell you things about meaning," he once said to me, "but it can't and wasn't intended to."[25]

Perhaps the most oversold metatheory was the appropriately named catastrophe theory, invented by the French mathematician René Thom in the 1960s. Thom developed the theory as a purely mathematical formalism, but he and others began to claim that it could provide insights into a broad range of phenomena that displayed discontinuous behavior. Thom's magnum opus was his 1972 book, *Structural Stability and Morphogenesis*, which received awestruck reviews in Europe and the United States. A reviewer in the *Times* of London declared that "it is impossible to give a brief description of the impact of this book. In one sense the only book with which it can be compared is Newton's *Principia*. Both lay out a new conceptual framework for the understanding of nature, and equally both go on to unbounded speculation."[26]

Thom's equations revealed how a seemingly ordered system could undergo abrupt, "catastrophic" shifts from one state to another. Thom and his followers suggested that these equations could help to explain not only such purely physical events as earthquakes, but also biological and social phenomena, such as the emergence of life, the metamorphosis of a caterpillar into a butterfly, and the collapse of civilizations. By the late 1970s, the counterattack had begun. Two mathematicians declared in *Nature* that catastrophe theory "is one of many attempts to deduce the world by thought alone." They called that "an appealing dream, but a dream that cannot come true." Other critics charged that Thom's work "provides no new information about anything" and is "exaggerated, not wholly honest."[27]

Chaos, as defined by James Yorke, exhibited this same boom–bust cycle. By 1991, at least one pioneer of chaos theory, the French mathematician David Ruelle, had begun to wonder whether his field had passed its peak. Ruelle invented the concept of strange attractors, mathematical objects that have fractal properties and can be used to describe the behavior of systems that never settle into a periodic pattern. In his book *Chance and Chaos*, Ruelle noted that chaos "has been invaded by swarms of people who are attracted by success, rather than the ideas involved. And this changes the intellectual atmosphere for the worse. . . . The physics of chaos, in spite of frequent triumphant announcements of 'novel' breakthroughs, has had a declining output of interesting discoveries. Hopefully, when the craze is over, a sober appraisal of the difficulties of the subject will result in a new wave of high-quality results."[28]

"More Is Different"

Even some researchers associated with the Santa Fe Institute seem to doubt that science can achieve the kind of transcendent, unified theory of complex phenomena that John Holland, Per Bak, and Stuart Kauffman all dream of. One skeptic is Philip Anderson, a notoriously hard-nosed physicist who won a Nobel Prize in 1977 for his work on superconductivity and other antics of condensed matter and was one of the founders of the Santa Fe Institute. Anderson was a pioneer of antireductionism. In "More Is Different," an essay published in *Science* in 1972, Anderson contended that particle physics, and indeed all reductionist approaches, have a limited ability to explain the world. Reality has a hierarchical structure, Anderson argued, with each level independent, to some degree, of the levels above and below. "At each stage, entirely new laws, concepts and generalizations are necessary, requiring inspiration and creativity to just as great a degree as in the previous one," Anderson noted. "Psychology is not applied biology, nor is biology applied chemistry."[29]

"More Is Different" became a rallying cry for the chaos and complexity movements. Ironically, the principle suggests that these so-called antireductionist efforts may never culminate in a unified theory of complex, chaotic systems, one that illuminates everything from immune systems to economies, as chaoplexologists such as Bak have suggested. (The principle also suggests that the attempt of Roger Penrose to explain the mind in terms of quasi-quantum mechanics was misguided.) Anderson acknowledged as much when I visited him at Princeton University, his home base. "I don't think there is a theory of everything," he said. "I think that there are basic principles that have very wide generality," such as quantum mechanics, statistical mechanics, thermodynamics, and symmetry breaking. "But you mustn't give in to the temptation that when you have a good general principle at one level that it's going to work at all levels." (Of quantum mechanics, Anderson said, "There seems to me to be no possible modification of that in the foreseeable future.") Anderson agreed with the evolutionary biologist Steven Jay Gould that life is shaped less by deterministic laws than by contingent, unpredictable circumstances. "I guess the prejudice I'm trying to express is a prejudice in favor of natural history," Anderson elaborated.

Anderson did not share the faith of some of his colleagues at the Santa Fe Institute in the power of computer models to illuminate complex systems. "Since I know a little bit about global economic models," he explained, "I know they don't work! I always wonder whether global climate models or oceanic circulation models or things like that are as full of phony statistics and phony measurements." Making simulations more detailed and realistic is not necessarily the solution, Anderson noted. It is possible for a computer to simulate the phase transition of a liquid into a glass, for example, "but have you learned anything? Do you understand it any better than you did before? Why not just take a piece of glass and say it's going through the glass transition? Why do you have to look at a computer go through the glass transition? So that's the *reductio ad absurdum*. At some point the computer's not telling you what the system itself is doing."

And yet, I said, there seemed to be this abiding faith among some of his colleagues that they would someday find a theory that would dispel all mysteries. "Yeah," Anderson said, shaking his head. Abruptly, he threw his arms in the air and cried out, like a born-again parishioner: "I've finally seen the light! I understand everything!" He lowered his arms and smiled ruefully. "You *never* understand everything," he said. "When one understands everything, one has gone crazy."

The Quark Master Rules Out "Something Else"

An even more improbable leader of the Santa Fe Institute is Murray Gell-Mann. Gell-Mann is a master reductionist. He won a Nobel Prize in 1969 for discovering a unifying order beneath the alarmingly diverse particles streaming from accelerators. He called his particle-classification system the Eight-fold way, after the Buddhist road to wisdom. (The name was meant to be a joke, Gell-Mann has often pointed out, not an indication that he was one of these flaky New Age types who thought physics and Eastern mysticism had something in common.) He showed the same flair for discerning unity in complexity—and for coining terms—when he proposed that neutrons, protons, and a host of other shorter-lived particles were all made of triplets of more fundamental particles called quarks. Gell-Mann's quark theory has been amply demonstrated in accelerators, and it remains a cornerstone of the standard model of particle physics.

Gell-Mann is fond of recalling how he found the neologism *quark* while perusing James Joyce's *Finnegans Wake*. (The passage reads, "Three quarks for Muster Mark!") This anecdote serves notice that Gell-Mann's intellect is far too powerful and restless to be satisfied by particle physics alone. According to a "personal statement" that he distributes to reporters, his interests include not only particle physics and modern literature, but also cosmology, nuclear arms–control policy, natural history, human history, population growth, sustainable human development, archaeology, and the evolution of language. Gell-Mann seems to have at least some familiarity with most of the major languages of the world and with many dialects; he enjoys telling people about the etymology and correct native pronunciation of their names. He was one of the first major scientists to climb aboard the complexity bandwagon. He helped to found the Santa Fe Institute, and he became its first full-time professor in 1993. (He had spent almost 40 years before that as a professor at the California Institute of Technology.)

Gell-Mann is unquestionably one of this century's most brilliant scientists. (His literary agent, John Brockman, said that Gell-Mann "has five brains, and each one is smarter than yours.")[30] He may also be one of the most annoying. Virtually everyone who knows Gell-Mann has a story about his compulsion to tout his own talents and to belittle those of others. He displayed this trait almost immediately after we first met in 1991 in a

New York City restaurant, several hours before he was to catch a flight to California. Gell-Mann is a small man with large black glasses, short white hair, and a skeptical squint. I had barely sat down when he began to tell me—as I set out my tape recorder and yellow pad—that science writers were "ignoramuses" and a "terrible breed" who invariably got things wrong; only scientists were really qualified to present their work to the masses. As time went on, I felt less offended, since it became clear that Gell-Mann held most of his scientific colleagues in contempt as well. After a series of particularly demeaning comments about some of his fellow physicists, Gell-Mann added, "I don't want to be quoted insulting people. It's not nice. Some of these people are my friends."

To stretch out the interview, I had arranged for a limousine to take us to the airport together. Once there, I accompanied Gell-Mann as he checked his bags and went to the first-class lounge. He began to fret that he did not have enough money to take a taxi home after he arrived in California. (Gell-Mann had not yet moved permanently to Santa Fe.) If I could lend him some money, he would write me a check. I gave him $40. As Gell-Mann handed me a check, he suggested that I consider not cashing it, since his signature would probably be quite valuable someday. (I cashed the check, but I kept a photocopy.)[31]

My suspicion is that Gell-Mann doubts that his colleagues in Santa Fe will discover anything truly profound, anything approaching, say, Gell-Mann's own quark theory. However, if a miracle occurs, and the chaoplexologists somehow manage to accomplish something important, Gell-Mann wants to be able to share the glory. His career would therefore encompass the whole range of modern science, from particle physics on up to chaos and complexity.

For a putative leader of chaoplexity, Gell-Mann espouses a worldview remarkably similar to that of the arch-reductionist Steven Weinberg—although of course Gell-Mann does not put it that way. "I have no idea what Weinberg said in his book," Gell-Mann said when I asked him during an interview in Santa Fe in 1995 if he agreed with what Weinberg had said in *Dreams of a Final Theory* about reductionism. "But if you read *mine* you saw what *I* said about it." Gell-Mann then went on to repeat some of the major themes of his 1994 book, *The Quark and the Jaguar*.[32] To Gell-Mann, science forms a hierarchy. At the top are those theories that apply everywhere in the known universe, such as the second law of thermodynamics and his own quark theory. Other laws, such as those related to genetic transmission, apply only here on earth, and the phenomena they describe entail a great deal of randomness and historical circumstance.

"With biological evolution we see a gigantic amount of history enters, huge numbers of accidents that could have gone different ways and produced different life-forms than we have on the earth, constrained of course by selection pressures. Then we get to human beings and the characteristics of human beings are determined by huge amounts of history. But still, there's clear determination from the fundamental laws and from history, or fundamental laws and specific circumstances."

Gell-Mann's reductionist predilections can be seen in his attempts to get his colleagues at the Santa Fe Institute to substitute his own neologism, *plectics*, for complexity. "The word is based on the Indo-European word *plec*, which is the basis of both simplicity and complexity. So in plectics we try to understand the relation between the simple and the complex, and in particular how we get from the simple fundamental laws that govern the behavior of all matter to the complex fabric that we see around us," he said. "We're trying to make theories of how this process works, in general and also special cases, and how those special cases relate to the general situation." (Unlike *quark*, *plectics* has never caught on. I have never heard anyone other than Gell-Mann use the term—except to deride Gell-Mann's fondness for it.)

Gell-Mann rejected the possibility that his colleagues would discover a single theory that embraced all complex adaptive systems. "There are huge differences among these systems, based on silicon, based on protoplasm, and so on. It's not the same." I asked Gell-Mann if he agreed with the "More Is Different" principle set forth by his colleague Philip Anderson. "I have no idea what he said," Gell-Mann replied disdainfully. I explained Anderson's idea that reductionist theories have limited explanatory power; one cannot go back up the chain of explanation from particle physics to biology. "You can! You can!" Gell-Mann exclaimed. "Did you read what I wrote about this? I devoted two or three chapters to this!"

Gell-Mann said that in principle one can go back up the chain of explanation, but in practice one often cannot, because biological phenomena stem from so many random, historical, contingent circumstances. That is not to say that biological phenomena are ruled by some mysterious laws of their own that act independently of the laws of physics. The whole point of the doctrine of emergence is that "we don't need *something else* in order to get *something else*," Gell-Mann said. "And when you look at the world that way, it just falls into place! You're not tortured by these strange questions any more!"

Gell-Mann thus rejected the possibility—raised by Stuart Kauffman and others—that there might be a still-undiscovered law of nature that explains why the universe has generated so much order in spite of the supposedly universal drift toward disorder decreed by the second law of thermodynamics. This issue, too, was settled, Gell-Mann replied. The universe began in a wound-up state far from thermal equilibrium. As the universe winds down, entropy increases, on average, throughout the system, but there can be many local violations of that tendency. "It's a tendency, and there are lots and lots of eddies in that process," he said. "That's *very* different from saying complexity increases! The envelope of complexity grows, expands. It's obvious from these other considerations it doesn't need another new law, however!"

The universe does create what Gell-Mann calls frozen accidents—galaxies, stars, planets, stones, trees—complex structures that serve as a foundation for the emergence of still more complex structures. "As a general rule, more complex life-forms emerge, more complex computer programs, more complex astronomical objects emerge in the course of nonadaptive stellar and galactic evolution and so on. But! If we look very very very far into the future, maybe it won't be true any more!" Eons from now the era of complexity could end, and the universe could degenerate into "photons and neutrinos and junk like that and not a lot of individuality." The second law would get us after all.

"What I'm trying to oppose is a certain tendency toward obscurantism and mystification," Gell-Mann continued. He emphasized that there was much to be understood about complex systems; that was why he helped to found the Santa Fe Institute. "There's a huge amount of wonderful research going on. What I say is that there is no evidence that we need—I don't know how else to say it—*something else!*" Gell-Mann, as he spoke, wore a huge sardonic grin, as if he could scarcely contain his amusement at the foolishness of those who might disagree with him.

Gell-Mann noted that "the last refuge of the obscurantists and mystifiers is self-awareness, consciousness." Humans are obviously more intelligent and self-aware than other animals, but they are not qualitatively different.

"*Again*, it's a phenomenon that appears at a certain level of complexity and presumably emerges from the fundamental laws plus an awful lot of historical circumstances. Roger Penrose has written two foolish books based on the long-discredited fallacy that Gödel's theorem has something to do with consciousness requiring"—pause—"*something else*."

If scientists did discover a new fundamental law, Gell-Mann said, they would do so by forging further into the microrealm, in the direction of superstring theory. Gell-Mann felt that superstring theory would probably be confirmed as the final, fundamental theory of physics early in the next millennium. But would such a far-fetched theory, with all its extra dimensions, ever really be accepted? I asked. Gell-Mann stared at me as if I had expressed a belief in reincarnation. "You're looking at science in this weird way, as if it were a matter of an opinion poll," he said. "The world is a certain way, and opinion polls have nothing to do with it! They do exert pressures on the scientific enterprise, but the ultimate selection pressure comes from comparison with the world." What about quantum mechanics? Would we be stuck with its strangeness? "I don't think there's anything strange about it! It's just quantum mechanics! Acting like quantum mechanics! That's all it does!" To Gell-Mann, the world made perfect sense. He already had *The Answer*.

Is science finite or infinite? For once, Gell-Mann did not have a prepackaged answer. "That's a very difficult question," he replied soberly. "I can't say." His view of how complexity emerges from fundamental laws, he said, "still leaves open the question of whether the whole scientific enterprise is open-ended. After all, the scientific enterprise can also concern itself with all kinds of details." Details.

One of the things that makes Gell-Mann so insufferable is that he is almost always right. His assertion that Kauffman, Bak, Penrose, and others will fail to find *something else* just beyond the horizon of current science—something that can explain better than current science can the mystery of life and of human consciousness and of existence itself—will probably prove to be correct. Gell-Mann may err—dare one say it?—only in thinking that superstring theory, with all its extra dimensions and its infinitesimal loops, will ever become an accepted part of the foundation of physics.

Ilya Prigogine and the End of Certainty

In 1994, Arturo Escobar, an anthropologist at Smith College, wrote an essay in the journal *Current Anthropology* about some of the new concepts and metaphors emerging from modern science and technology. He noted that chaos and complexity offered different visions of the world than did traditional science; they emphasized "fluidity, multiplicity, plurality, connectedness, segmentarity, heterogeneity, resilience; not 'science' but knowledges of the concrete and the local, not laws but knowledge of the problems and the self-organizing dynamics of nonorganic, organic and social phenomena." Note the quotation marks around the word *science*.[33]

It is not only postmodernists such as Escobar who view chaos and complexity as what I would call ironic endeavors. The artificial-life maven Christopher Langton clearly was putting forward this same idea when he foresaw more "poetry" in the future of science. Langton's ideas, in turn, echoed those espoused much earlier by the chemist Ilya Prigogine. In 1977, Prigogine won a Nobel Prize for studies of so-called dissipative systems, unusual mixtures of chemicals that never achieve equilibrium but keep fluctuating among multiple states. Upon these experiments, Prigogine, who fluctuates between institutes that he founded at the Free University in Belgium and the University of Texas at Austin, constructed a tower of ideas about self-organization, emergence, the links between order and disorder—in short, chaoplexity.

Prigogine's great obsession is time. For decades he complained that physics was not paying sufficient heed to the obvious fact that time proceeds only in one direction. In the early 1990s, Prigogine announced that he had forged a new theory of physics, one that finally did justice to the irreversible nature of reality. The probabilistic theory supposedly eliminated the philosophical paradoxes that had plagued quantum mechanics and reconciled it with classical mechanics, nonlinear dynamics, and thermodynamics. As a bonus, Prigogine declared, the theory would help to bridge the chasm between the sciences and the humanities and bring about the "reenchantment" of nature.

Prigogine has his fans, at least among nonscientists. Futurist Alvin Toffler, in the forward to Prigogine's 1984 book, *Order out of Chaos* (a bestseller in Europe), likened Prigogine to Newton and prophesied that the science of the Third Wave future would be Prigoginian.[34] But scientists familiar with Prigogine's work—including the many younger practitioners of chaos and complexity who have clearly borrowed his ideas and rhetoric—have little or nothing good to say about him. They accuse him of being arrogant and self-aggrandizing. They claim that he has made little or no concrete contribution to science; that he has merely recreated experiments by others and waxed philosophical about them; and that he had won a Nobel Prize for less cause than any other recipient.

These charges may well be true. But Prigogine may also have earned his colleagues' enmity by revealing the dirty secret of late-twentieth-century science, that it is, in a sense, digging its own grave. In *Order out of Chaos*, cowritten with Isabelle Stengers, Prigogine pointed out that the major discoveries of science in this century have proscribed the limits of science. "Demonstrations of impossibility, whether in relativity, quantum mechanics, or thermodynamics, have shown us that nature cannot be described 'from the outside,' as if by a spectator," Prigogine and Stengers stated. Modern science, with its probabilistic descriptions, also "leads to a kind of 'opacity' as compared to the transparency of classical thought."[35]

I met Prigogine in Austin in March 1995, a day after his return from a stint in Belgium. He showed no signs of jet lag. At age 78, he was exceedingly alert and energetic. Although short and compact, he possessed a regal bearing; he seemed not arrogant so much as calmly accepting of his own greatness. As I reviewed the issues I hoped to discuss with him, he nodded and muttered, "Yays, yays," with some impatience; he was eager, I soon realized, to begin enlightening me about the nature of things.

Shortly after we sat down, two researchers at the center joined us. A secretary told me later that they had been instructed to interrupt Prigogine if he proved too overwhelming for me and did not allow me to ask any questions. They did not fulfill this task. Prigogine, once started, was unstoppable. Words, sentences, paragraphs issued from him in a steady, implacable stream. His accent was almost parodically thick—it reminded me of Inspector Clouseau in *The Pink Panther*—and yet I had no difficulty understanding him.

He briefly recounted his youth. He was born in Russia in 1917, during the revolution, and his bourgeois family soon fled to Belgium. His interests were eclectic: he played the piano, studied literature, art, philosophy—and science, of course. He suspected that the turbulent setting of his youth had provoked his career-long fascination with time. "I may have been impressed with the fact that science had so little to say about time, about history, evolution, and that perhaps brought me to the problem of thermo-dynamics. Because in thermodynamics, the main quantity is entropy, and entropy means just evolution."

In the 1940s, Prigogine proposed that the increase of entropy decreed by the second law of thermodynamics need not always create disorder; in some systems, such as the churning chemical cells he studied in his labora-tory, entropic drift could generate striking patterns. He also began to realize that "structure is rooted in the direction of time, irreversible time, and the arrow of time is a very important element in the structure of the universe. Now this already was bringing me in a sense in conflict with great physicists like Einstein, who was saying that time is an illusion."

Most physicists, according to Prigogine, think irreversibility is an illu-sion stemming from the limits of their observations. "Now this I could never believe, because in a sense that seems to indicate that our measure-ments, or our approximations, introduce irreversibility in a time-reversible universe!" Prigogine exclaimed. "We are not the father of time. We are the children of time. We come from evolution. What we have to do is to include evolutionary patterns in our descriptions. What we need is a Darwinian view of physics, an evolutionary view of physics, a biological view of physics."

Prigogine and his colleagues had been developing just such a physics. As a result of this new model, Prigogine told me, physics would be reborn, contrary to the pessimistic predictions of reductionists such as Steven Weinberg (who worked in the same building as Prigogine). The new physics might also heal the great rift between science, which had always depicted nature as the outcome of deterministic laws, and the humanities, which emphasized human freedom and responsibility. "You cannot on one side believe that you are part of an automaton and on the other hand believe in humanism," Prigogine declared.

Of course, he emphasized, this unification was metaphorical rather than literal; it would not by any means help science to solve all its problems. "One should not exaggerate and dream about a unified theory which will include politics and economics and the immune system and physics and chemistry," Prigogine said, in a rebuke to researchers at the Santa Fe Institute and elsewhere who dreamed of just such a theory. "One should not think that progress in chemical nonequilibrium reactions will give you the key for human politics. Of course not! Of course not! But still, it brings in a unified element. It brings in the element of bifurcation, it brings in the idea of historical dimension, it brings in the idea of evolutionary patterns, which indeed you find on all levels. And in this sense it is a unifying element of our view of the universe."

Prigogine's secretary poked her head in at the door to remind him that she had made reservations for lunch at the faculty club at noon. After her third reminder—at five minutes after noon—Prigogine ended his peroration with a flourish and announced that it was time to go to lunch. At the faculty lounge, Prigogine and I were joined by a dozen or so other researchers employed at his center: Prigoginians. We assembled at a long rectangular table. Prigogine sat in the middle of one side, like Jesus at the Last Supper, and I sat beside him, like Judas, listening, along with everyone else, as he held forth.

Sporadically, Prigogine called on one of his disciples to say a word or two—enough to draw attention to the vast gap between his rhetorical powers and theirs. At one point he asked a tall, cadaverish man sitting across from me (whose equally cadaverish identical twin, eerily, was also at the table) to explain his nonlinear, probabilistic view of cosmology. The man dutifully unburdened himself, in a lugubrious eastern European accent, of an impenetrable monologue about bubbles and instabilities and quantum fluctuations. Prigogine quickly stepped in. The meaning of his colleague's work, he explained, was that there was no stable ground state, no equilibrium condition, of space-time; thus, there was no beginning to the cosmos and there could be no end. Phew.

Between nibbles of his fish, Prigogine reiterated his objections to determinism. (Earlier Prigogine had admitted that he had been strongly influenced by Karl Popper.) Descartes, Einstein, and the other great determinists were "all pessimistic people. They wanted to go to another world, a world of eternal beatitude." But a deterministic world would be not a utopia but a *dys*topia, Prigogine said. That was the message of Aldous Huxley in *Brave New World*, of George Orwell in *1984*, of Milan Kundera in *The Unbearable Lightness of Being*. When a state tries to suppress evolution, change, flux by brutal force, by violence, Prigogine explained, it destroys the meaning of life, it creates a society of "timeless robots."

On the other hand, a completely irrational, unpredictable world would also be terrifying. "What we have to find is a middle way, to find a probabilistic description which says something, not everything, and also not nothing." His view could provide a philosophical framework for understanding social phenomena, Prigogine said. But human behavior, he emphasized, could not be defined by any scientific, mathematical model. "In human life we have no simple basic equation! When you decide whether you take coffee or not, that's already a complicated decision. It depends on what day it is, whether you like coffee, and so on."

Prigogine had been building toward some great revelation, and now he finally divulged it. Chaos, instability, nonlinear dynamics, and related concepts had been warmly received not only by scientists but also by the lay public because society was itself in a state of flux. The public's faith in great unifying ideas, whether religious or political or artistic or scientific, was dwindling.

"Even people who are very Catholic are no longer so Catholic as their parents or grandparents were, probably. We are no more believing in Marxism or liberalism in the classical way. We are no more believing classical science." The same is true of the arts, music, literature; society has learned to accept a multiplicity of styles and worldviews. Humanity has arrived, Prigogine summarized, at "the end of certitude."

Prigogine paused, allowing us to ponder the magnitude of his announcement. I broke the hushed silence by pointing out that some people, such as religious fundamentalists, seemed to be clinging to certitude more fiercely than ever. Prigogine listened politely, then asserted that fundamentalists were merely exceptions to the rule. Abruptly, he fixed his gaze on a prim, blond-haired woman, the deputy director of his institute, sitting across the table from us. "What is your opinion?" he asked. "I agree completely," she replied. She hastily added, perhaps in response to the craven snickers of her colleagues, that fundamentalism "seemed to be a response to a frantic world."

Prigogine nodded paternally. He acknowledged that his assertions concerning the end of certitude had elicited "violent reactions" in the intellectual establishment. The *New York Times* had declined to review *Order out of Chaos* because, Prigogine had heard, the editors considered his discussion of the end of certitude "too dangerous." Prigogine understood such fears. "If science is not able to give certitude, what should you believe? I mean, before it was very easy. Either you believe in Jesus Christ, or you believe in

Newton. It was very simple. But now, as I say, if science is not giving you certitude but probabilities, then it's a dangerous book!"

Prigogine nonetheless thought his view did justice to the depthless mystery of the world, and of our own existence. That was what he meant by his phrase "the reenchantment of nature." After all, consider this lunch we were having right now. What theory could predict this! "The universe is a strange thing," Prigogine said, cranking the intensity of his voice up a notch. "I think we can all agree on that." As he swept the room with his serene yet feral gaze, his colleagues bobbed their heads and chuckled nervously. Their unease was well-founded. They had hitched their careers to a man who apparently believed that science—empirical, rigorous science, the kind of science that solves its problems, that renders the world comprehensible, that gets us somewhere—was over.

In return for certainty, Prigogine—like Christopher Langton, Stuart Kauffman, and other chaoplexologists who have clearly been influenced by his ideas—promises the "reenchantment of nature." (Per Bak, for all his hubris, at least eschews this pseudo-spiritual rhetoric.) What Prigogine apparently means by this statement is that vague, fuzzy, impotent theories are somehow more meaningful, more comforting, than the accurate, precise, powerful theories of Newton or Einstein or modern particle physicists. But why, one wonders, is an indeterministic, opaque universe any less cold, cruel, and frightening than a deterministic, transparent one? More specifically, how is the fact that the world unfolds according to a nonlinear, probabilistic dynamics supposed to console a Bosnian woman who has seen her only daughter raped and slaughtered?

Mitchell Feigenbaum and the Collapse of Chaos

It was a meeting with Mitchell Feigenbaum that finally convinced me that chaoplexity is a doomed enterprise. Feigenbaum was perhaps the most compelling character in Gleick's book *Chaos*—and in the field as a whole. Trained as a particle physicist, Feigenbaum became entranced with questions beyond the scope of that or any other field, questions about turbulence, chaos, and the relation between order and disorder. In the mid-1970s, when he was a young postdoc at Los Alamos National Laboratory, he discovered a hidden order, called period doubling, underlying the behavior of a wide variety of nonlinear mathematical systems. The period of a system is the time it takes to return to its original state. Feigenbaum found that the period of some nonlinear systems keeps doubling as they evolve and therefore rapidly approaches infinity (or eternity). Experiments confirmed that some simple real-world systems (although not as many as initially hoped) demonstrate period doubling. For example, as one gradually opens up a faucet, the water demonstrates period doubling as it progresses from a steady drip, drip, drip toward a turbulent gush. The mathematician David Ruelle has called period doubling a work of "particular beauty and significance" that "stands out in the theory of chaos."[36]

Feigenbaum, when I met him in March 1994 at Rockefeller University, where he has a spacious office overlooking Manhattan's East River, looked every bit the genius he was said to be. With his magnificent, oversized head and swept-back hair, he resembled Beethoven, though more handsome and less simian. Feigenbaum spoke clearly, precisely, with no accent, but with a strange kind of formality, as if English were a second language that he had mastered through sheer brilliance. (The voice of the superstring theorist Edward Witten has this same quality.) When amused, Feigenbaum did not smile so much as grimace: his already protuberant eyes bulged still farther from their sockets, and his lips peeled back to expose twin rows of brown, peglike teeth stained by countless filterless cigarettes and espressos (both of which he consumed during our meeting). His vocal cords, cured by decades of exposure to these toxins, yielded a voice as rich and resonant as a basso profundo's and a deep, villainous snicker.

Like many chaoplexologists, Feigenbaum could not resist ridiculing particle physicists for daring to think they could achieve a theory of everything. It is quite possible, he said, that particle physicists might one day develop a theory that adequately accounts for all of nature's fundamental forces, including gravity. But calling such a theory final would be something else again. "A lot of my colleagues like the idea of final theories because they're religious. And they use it as a replacement for God, which they don't believe in. But they just created a substitute."

A unified theory of physics would obviously not answer all questions, Feigenbaum said. "If you really believe that this is a path of understanding the world, I can ask immediately: how do I write down in this formalism what you look like, with all the hairs on your head?" He stared at me until my scalp prickled. "Now, one answer is, 'That's not an interesting problem.' " Against my will, I felt slightly offended. "Another answer is, 'Well, it's okay, but we can't do it.' The right answer is obviously an alloy of those two complements. We have very few tools. We can't solve problems like that."

Moreover, particle physicists are overly concerned with finding theories that are merely true, in the sense that they account for available data; the goal of science should be to generate "thoughts in your head" that "stand a high chance of being new or exciting," Feigenbaum explained. "*That's* the desideratum." He added: "There isn't any security by knowing that something is true, at least as far as I'm concerned. I'm thoroughly indifferent to that. I like to know that I have a way of thinking about things." I began to suspect that Feigenbaum, like David Bohm, had the soul of an artist, a poet, even a mystic: he sought not truth, but revelation.

Feigenbaum noted that the methodology of particle physics—and physics generally—had been to try to look at the simplest possible aspects of reality, "where everything has been stripped away." The most extreme reductionists had suggested that looking at more complex phenomena was merely "engineering." But as a result of advances in chaos and complexity, he said, "some of these things that one relegated to engineering are now regarded as reasonable questions to ask from a more theoretical viewpoint. Not just to get the right answer but to understand something about how they work. And that you can even make sense out of that last comment flies in the face of what it means for a theory to be finished."

On the other hand, chaos, too, had generated too much hype. "It's a fraud to have named the subject 'chaos,' " he said. "Imagine one of my [particle-physicist] colleagues has gone to a party and meets someone and the person is all bubbling over about chaos and tells him that this reductionist stuff is all bullshit. Well, it's infuriating, because it's completely stupid what the person has been told," Feigenbaum said. "I think it's regrettable that people are sloppy, and they end up serving as representatives."

Some of his colleagues at the Santa Fe Institute, Feigenbaum added, also had too naive a faith in the power of computers. "The proof is in the pudding," he said, and paused, as if considering how to proceed diplomatically. "It's very hard to see things in numerical experiments. That is, people want to have fancier and fancier computers to simulate fluids. There is something to be learned in simulating fluids, but unless you know what you're looking for, you're not going to see anything. Because after all, if I just look out the window, there's an overwhelmingly better simulation than I could ever do on a computer."

He nodded toward his window, beyond which the leaden East River flowed. "I can't interrogate it quite as sharply, but there's so much stuff in that numerical simulation that if I don't know what to interrogate it about, I will have learned nothing." For these reasons much of the recent work on nonlinear phenomena "has not led to answers. The reason for that is, these are truly hard problems, and one doesn't have any tools. And the job should really be to do those insightful calculations which require some piece of faith and good luck as well. People don't know how to begin doing these problems."

I admitted that I was often confused by the rhetoric of people in chaos and complexity. Sometimes they seemed to be delineating the limits of science—such as the butterfly effect—and sometimes they implied that they could transcend those limits. "We are building tools!" Feigenbaum cried. "We don't know how to do these problems. They are truly hard. Every now and then we get a little pocket where we know how to do it, and then we try to puff it out as far as it can go. And when it reaches the border of where it is going, then people wallow for a while, and then they stop doing it. And then one waits for some new piece of insight. But it is literally the business of enlarging the borders of what falls under the suzerainty of science. It is *not* being done from an engineering viewpoint. It isn't just to give you the answer to some approximation."

"I want to know *why*," he continued, still staring at me hard. "*Why* does the thing do this?" Was it possible that this enterprise could, well, fail? "Of course!" Feigenbaum bellowed, and he laughed maniacally. He confessed that he had been stymied himself of late. Up through the late 1980s he had sought to refine a method for describing how a fractal object, such as a cloud, might evolve over time when perturbed by various forces. He wrote two long papers on the topic that were published in 1988 and 1989 in a relatively obscure physics journal.[37] "I have no idea how well they've been read," Feigenbaum said defiantly. "In fact, I've never been able to give a talk on them." The problem, he suggested, might be that no one could understand what he was getting at. (Feigenbaum was renowned for obscurity as well as for brilliance.) Since then, he added, "I haven't had a further better idea to know how to proceed in this."

In the meantime, Feigenbaum had turned to applied science. Engineering. He had helped a map-making company develop software for automatically constructing maps with minimal spatial distortion and maximum aesthetic appeal. He belonged to a committee that was redesigning U.S. currency to make it less susceptible to counterfeiting. (Feigenbaum came up with the idea of using fractal patterns that blur when photocopied.) I noted that these sounded like what would be, for most scientists, fascinating and worthy projects. But people familiar with Feigenbaum's former incarnation as a leader of chaos theory, if they heard he now worked on maps and currency, might think . . .

"He's not doing serious things any more," Feigenbaum said quietly, as if to himself. Not only that, I added. What people might think was that if someone who was arguably the most gifted explorer of chaos could not proceed any further, then perhaps the field had run its course. "There's *some* truth to that," he replied. He acknowledged that he hadn't really had any good ideas about how to extend chaos theory since 1989. "One is on the lookout for things that are substantial, and at the moment . . ." He paused. "I don't have a thought. I don't know . . ." He turned his large, luminous eyes once again toward the river beyond his window, as if seeking a sign.

Feeling somewhat guilty, I told Feigenbaum that I would love to see his last papers on chaos. Did he have any reprints? In response, Feigenbaum thrust himself from his chair and careened wildly toward a row of filing cabinets on the far side of his office. En route, he cracked his shin against a low-lying coffee table. Wincing, teeth clenched, Feigenbaum limped onward, wounded by his collision with the world. The scene was a grotesque inversion of Samuel Johnson's famous stone-kicking episode. The suddenly malevolent-looking coffee table seemed to be gloating: "I refute Feigenbaum thus."

Making Metaphors

The fields of chaos, complexity, and artificial life will continue. Certain practitioners will be content to play in the realm of pure mathematics and theoretical computer science. Others, the majority, will develop new mathematical and computational techniques for engineering purposes. They will make incremental advances, such as extending the range of weather forecasts or improving the ability of engineers to simulate the performance of jets or other complex technologies. But they will not achieve any great insights into nature—certainly none comparable to Darwin's theory of evolution or quantum mechanics. They will not force any significant revisions in our map of reality or our narrative of creation. They will not find what Murray Gell-Mann calls "something else."

So far, chaoplexologists have created some potent metaphors: the butterfly effect, fractals, artificial life, the edge of chaos, self-organized criticality. But they have not told us anything about the world that is both concrete and truly surprising, either in a negative or in a positive sense. They have slightly extended the borders of knowledge in certain areas, and they have more sharply delineated the boundaries of knowledge elsewhere.

Computer simulations represent a kind of metareality within which we can play with and even—to a limited degree—test scientific theories, but they are not reality itself (although many aficionados have lost sight of that distinction). Moreover, by giving scientists more power to manipulate different symbols in different ways to simulate a natural phenomenon, computers may undermine scientists' faith that their theories are not only true but *True*, exclusively and absolutely true. Computers may, if anything, hasten the end of empirical science. Christopher Langton was right: there is something more like poetry in the future of science.

CHAPTER NINE

The End of Limitology

Just as lovers begin talking about their relationship only when it sours, so will scientists become more self-conscious and doubtful as their efforts yield diminishing returns. Science will follow the path already trodden by literature, art, music, philosophy. It will become more introspective, subjective, diffuse, obsessed with its own methods. In the spring of 1994 I saw the future of science in microcosm when I sat in on a workshop at the Santa Fe Institute titled "The Limits to Scientific Knowledge." During the three-day meeting, a score of thinkers, including mathematicians, physicists, biologists, and economists, pondered whether there were limits to science, and, if so, whether science could know them. The meeting was organized by two people associated with the Santa Fe Institute: John Casti, a mathematician who has written numerous popular books on science and mathematics, and Joseph Traub, a theoretical computer scientist, who is a professor at Columbia University.[1]

I had come to the meeting in large part to meet Gregory Chaitin, a mathematician and computer scientist at IBM who had devoted himself since the early 1960s to exploring and extending Gödel's theorem through what he called algorithmic information theory. Chaitin had come close to proving, as far as I could tell, that a mathematical theory of complexity was not possible. Before meeting Chaitin, I had pictured him as a gnarled, sour-looking man with hairy ears and an eastern European accent; after all, a kind of old-world philosophical angst suffused his research on the limits of mathematics. But Chaitin in no way resembled my internal model. Stout, bald, and boyish, he wore neobeatnik attire: baggy white pants with an elastic waistband, black T-shirt adorned with a Matisse sketch, sandals over socks. He was younger than I expected; I learned later that his first paper had been published when he was only 18, in 1965. His hyperactivity made him seem younger still. His speech was invariably either accelerating, as he became carried away by his words, or decelerating, perhaps as he realized he was approaching the limits of human comprehension and ought to slow down. Plots of the velocity and volume of his speech would form overlapping sine waves. Struggling to articulate an idea, he squeezed his eyes shut and, with an agonized grimace, tipped his head forward, as if trying to dislodge the words from his sticky brain.[2]

The participants sat around a long, rectangular table in a long, rectangular room with a chalkboard at one end. Casti opened the meeting by asking, "Is the real world too complex for us to understand?" Kurt Gödel's incompleteness theorem, Casti noted, implied that some mathematical descriptions would always be incomplete; some aspects of the world would always resist description. Alan Turing, similarly, showed that many mathematical propositions are "undecidable"; that is, one cannot determine whether the propositions are true or false in a finite amount of time. Traub tried to rephrase Casti's question in a more positive light: Can we know what we cannot know? Can we *prove* there are limits to science, in the same way that Gödel and Turing proved there are limits to mathematics and computation?

The only way to construct such a proof, announced E. Atlee Jackson, a physicist from the University of Illinois, would be to construct a formal representation of science. To show how difficult that task would be, Jackson leaped up to the chalkboard and scribbled an extremely complicated flowchart that supposedly represented science. When his audience stared at him blankly, Jackson fell back on aphorisms. To determine whether science has limits, he said, you have to define science, and as soon as you define science you impose a limit on it. On the other hand, he added, "I can't define my wife, but I can recognize her." Rewarded with polite chuckles, Jackson retreated to his seat.

The antichaos theorist Stuart Kauffman kept slipping into the meeting, delivering Zen-like minilectures, and then slipping out again. During one appearance, he reminded us that our very survival depends on our ability to classify the world. But the world doesn't come already packaged into premade categories. We can "carve it up," or classify it, in many ways. In order to classify phenomena, moreover, we must throw some information away. Kauffman concluded with this incantation: "To be is to classify is to act, all of which means throwing away information. So just the act of knowing requires ignorance." His audience looked simultaneously impressed and annoyed.

Ralph Gomory then said a few words. A former vice president of research at IBM, Gomory headed the Sloan Foundation, a philanthropic organization that sponsored science-related projects, including the Santa Fe workshop. When listening to someone else speak, and even when speaking himself, Gomory wore an expression of deep incredulity. He tilted his head forward, as if peering over invisible bifocals, while knitting his thick black eyebrows together and wrinkling his brow.

Gomory explained that he had decided to support the workshop because he had long felt that the educational system placed too much emphasis on what was known and too little on what was unknown or even unknowable. Most people aren't even aware of how little is known, Gomory said, because the educational system presents such a seamless, noncontradictory view of reality. Everything we know about the ancient Persian Wars, for example, derives from a single source, Herodotus. How do we know whether Herodotus was an accurate reporter? Maybe he had incomplete or inaccurate information! Maybe he was biased or making things up! We will never know!

Later Gomory remarked that a Martian, by observing humans playing chess, might be able to deduce the rules correctly. But could the Martian ever be sure that those were the correct rules, or the only rules? Everyone pondered Gomory's riddle for a moment. Then Kauffman speculated on how Wittgenstein might have responded to it. Wittgenstein would have "suffered egregiously," Kauffman said, over the possibility that the chess players might make a move—deliberately or not—that broke the rules. After all, how could the Martian tell if the move was just a mistake or the result of another rule? "Do you get this?" Kauffman queried Gomory.

"I don't know who Wittgenstein is, for starters," Gomory replied irritably.

Kauffman raised his eyebrows. "He was a *very* famous philosopher."

He and Gomory stared at each other until someone said, "Let's leave Wittgenstein out of this."

Patrick Suppes, a philosopher from Stanford, kept interrupting the discussion to point out that Kant, in his discussion of antinomies, anticipated virtually all the problems they were wrestling with at the workshop. Finally, when Suppes brought up yet another antinomy, someone shouted, "No more Kant!" Suppes protested that there was just one more antinomy he wanted to mention that was really very important, but his colleagues shouted him down. (No doubt they did not want to be reminded that for the most part they were merely restating, with newfangled jargon and metaphors, arguments set forth long ago not only by Kant but even by the ancient Greeks.)

Chaitin, spitting out words like a machine gun, dragged the conversation back to Gödel. The incompleteness theorem, Chaitin asserted, far from being a paradoxical curiosity with little relevance to the progress of mathematics or science, as some mathematicians like to believe, is only one of a set of profound problems posed by mathematics. "Some people dismiss Gödel's results as being bizarre, pathological, deriving from a self-referential paradox," Chaitin said. "Gödel himself sometimes worried that this was just a paradox created by our use of words. Now, incompleteness seems so natural, you can ask how we mathematicians can do anything!"

Chaitin's own work on algorithmic information theory suggested that as mathematicians address increasingly complex problems, they will have to keep adding to their base of axioms; to know more, in other words, one must assume more. As a result, Chaitin contended, mathematics is bound to become an increasingly experimental science with less of a claim to absolute truth. Chaitin had also established that just as nature seems to harbor fundamental uncertainty and randomness, so does mathematics. He had recently found an algebraic equation that might have an infinite or finite number of solutions, depending on the value of the variables in the equation.

"Normally you assume that if people think something is true, it's true for a reason. In mathematics a reason is called a proof, and the job of a mathematician is to find the proof, the reasons, deductions from axioms or accepted principles. Now what I found is mathematical truths that are true for no reason at all. They are true accidentally or at random. And that's why we will never find the truth: because there is no truth, there is no reason these are true."

Chaitin had also proved that one can never determine whether any given computer program is the most succinct possible method of solving a problem; it is always possible that more concise programs exist. (The implication of this finding, others have pointed out, is that physicists can never be sure that they have found a truly final theory, one that represents the most compact possible description of nature.) Chaitin obviously reveled in being the bearer of such dire tidings. He seemed intoxicated at the thought that he was tearing down the temple of mathematics and science.

Casti retorted that mathematicians might be able to avoid Gödel effects by employing simple formal systems—such as an arithmetic consisting solely of addition and subtraction (and not multiplication and division). Nondeductive systems of reasoning, Casti added, might also sidestep the problem; Gödel's theorem might turn out to be a red herring when it came to natural science.

Francisco Antonio "Chico" Doria, a Brazilian mathematician, also found Chaitin's analysis too pessimistic. The mathematical hurdles identified by Gödel, Doria contended, far from bringing mathematics to an end, could enrich it. For example, Doria suggested, when mathematicians encounter an apparently undecidable proposition, they can create two new branches of mathematics, one that assumes the proposition is true, and one that assumes it is false. "Instead of a limit of knowledge," Doria concluded, "we may have a wealth of knowledge."

Listening to Doria, Chaitin rolled his eyes. Suppes also seemed doubtful. Arbitrarily assuming that undecidable mathematical statements are true or false, Suppes drawled, has "all the advantages of theft over honest toil." He attributed his quip to someone famous.

The conversation kept veering—as if to a strange attractor—to one of the favorite topics of philosophically inclined mathematicians and physicists: the continuum problem. Is reality smooth or lumpy? Analog or digital? Is the world best described by the so-called real numbers, which can be diced into infinitely fine gradations, or by whole numbers? Physicists from Newton through Einstein have relied on real numbers. But quantum mechanics suggests that matter and energy and perhaps even time and space (at extremely small scales) come in discrete, indivisible lumps. Computers, too, represent everything as integers: ones and zeros.

Chaitin denounced real numbers as "nonsense." Their precision is a sham, given the noisiness, the fuzziness, of the world. "Physicists know that every equation is a lie," he declared.

Someone parried with a quote from Picasso: "Art is a lie that helps us see the truth."

Of *course* real numbers are abstractions, Traub chimed in, but they are very powerful, effective abstractions. Of *course* there is always noise, but there are ways to deal with noise in a real-number system. A mathematical model captures the essence of something. No one pretends it captures the *entire* phenomenon.

Suppes strode to the chalkboard and scribbled some equations that, he implied, might eliminate the continuum problem once and for all. His audience looked unimpressed. (This, I thought, is the major problem of philosophy: no one really *wants* to see philosophical problems solved, because then they will have nothing to talk about.)

Other participants noted that scientists face obstacles to knowledge much less abstract than incompleteness, undecidability, the continuum, and so on. One was Piet Hut, a Dutch astrophysicist from the Institute for Advanced Study. He said that with the help of powerful statistical methods and computers, he and his fellow astrophysicists had learned how to overcome the infamous N-body problem, which holds that it is impossible to predict the course of three or more gravitationally interacting bodies. Computers could now simulate the evolution of whole galaxies, containing billions of stars, and even clusters of galaxies.

But, Hut added, astronomers face other limits that seem insurmountable. They have only one universe to study, so they cannot do controlled experiments on it. Cosmologists can only trace the history of the universe so far back, and they can never know what preceded the big bang or what exists beyond the borders of our universe, if anything. Moreover, particle physicists may have a hard time testing theories (such as those involving superstrings) that combine gravity and all the other forces of nature, because the effects only become apparent at distance scales and energies that are beyond the range of any conceivable accelerator.

A similarly pessimistic note was sounded by Rolf Landauer, a physicist at IBM and a pioneer in the study of the physical limits of computation. Landauer spoke with a German-accented growl that sharpened the edge of his sardonic sense of humor. When one speaker kept standing in the way of his own viewgraphs, Landauer snapped, "Your talk may be transparent, but you are not!"

Landauer contended that scientists could not count on computers to keep growing in power indefinitely. He granted that many of the supposed physical constraints once thought to be imposed on computation by the second law of thermodynamics or quantum mechanics had been shown to be spurious. On the other hand, the costs of computer-manufacturing plants were rising so fast that they threatened to bring to a halt the decades-long decline in computation prices. Landauer also doubted whether computer designers would soon harness exotic quantum effects such as superposition—the ability of a quantum entity to exist in more than one state at the same time—and thereby transcend the capabilities of current computers, as some theorists had proposed. Such systems would be so sensitive to minute quantum-level disruptions that they would be effectively useless, Landauer argued.

Brian Arthur, a ruddy-faced economist at the Santa Fe Institute who spoke with a lilting Irish accent, steered the conversation toward the limits of economics. In trying to predict how the stock market will perform, he said, an investor must make guesses about how others will guess about how others will guess—and so on ad infinitum. The economic realm is inherently subjective, psychological, and hence unpredictable; indeterminacy "percolates through the system." As soon as economists try to simplify their models—by assuming that investors can have perfect knowledge of the market or that prices represent some true value—the models become unrealistic; two economists gifted with infinite intelligence will come to different conclusions about the same system. All economists can really do is say, "Well, it could be this, it could be that." On the other hand, Arthur added, "if you've made money in the markets all the economists will listen to you."

Kauffman then repeated what Arthur had just said but in a more abstract way. People are "agents" who must continually adjust their "internal models" in response to the perceived adjustments of the internal models of other agents, thus creating a "complex, coadapting landscape."

Landauer, scowling, interjected that there were much more obvious reasons that economic phenomena were impossible to predict than these subjective factors. AIDS, third-world wars, even the diarrhea of the chief analyst of a large mutual fund can have a profound effect on the economy, he said. What model can possibly predict such events?

Roger Shepard, a psychologist from Stanford who had been listening in silence, finally piped up. Shepard seemed faintly melancholy. His apparent mood may have been an illusion engendered by his droopy, ivory-hued mustache—or a very real by-product of his obsession with unanswerable questions. Shepard admitted he had come here in part to learn whether scientific or mathematical truths were discovered or invented. He had also been thinking a good deal lately about where scientific knowledge really existed and had concluded that it could not exist independently of the human mind. A textbook on physics, without a human to read it, is just paper and ink spots. But that raised what Shepard considered to be a disturbing issue. Science appears to be getting more and more complicated and thus more and more difficult to understand. It seems quite possible that in the future some scientific theories, such as a theory of the human mind, will be too complex for even the most brilliant scientist to understand. "Maybe I'm old-fashioned," Shepard said, but if a theory is so complicated that no single person can understand it, what satisfaction can we take in it?

Traub, too, was troubled by this issue. We humans may believe Occam's razor—which holds that the best theories are the simplest—because these are the only theories our puny brains can comprehend. But maybe computers won't be subject to this limitation, Traub added. Maybe computers will be the scientists of the future.

In biology, someone remarked gloomily, "Occam's razor cuts your throat."

Gomory noted that the task of science was to find those niches of reality that lend themselves to understanding, given that the world is basically unintelligible. One way to make the world more comprehensible, Gomory suggested, is to make it more artificial, since artificial systems tend to be more comprehensible and predictable than natural ones. For example, to make weather forecasting easier, society might encase the world in a transparent dome.

Everyone stared at Gomory for a moment. Then Traub remarked, "I think what Ralph is saying is that it's easier to create the future than to predict it."

As the meeting progressed, Otto Rossler made more and more sense. Or was everybody else making less? Rossler was a theoretical biochemist and chaos theorist from the University of Tübingen, Germany, who in the mid-1970s had discovered a mathematical monster called the Rossler attractor. His white hair appeared permanently disheveled, as though he had just awakened from a trance. He had the exaggerated features of a marionette: startled eyes, a protuberant lower lip, and a bulbous chin framed by deep, vertical creases. Neither I nor, I suspected, anyone else could quite understand him, but everyone leaned toward him when he issued his whispery, stammered pronouncements, as if he were an oracle.

Rossler saw two primary limits to knowledge. One was inaccessibility. We can never be sure about the origin of the universe, for example, because it is so distant from us both in space and in time. The other limit, distortion, was much worse. The world can deceive us into thinking we understand it when actually we do not. If we could stand outside the universe, Rossler suggested, we would know the limits to our knowledge; but we are trapped inside the universe, and so our knowledge of our own limits must remain incomplete.

Rossler raised some questions that he said were first posed in the eighteenth century by a physicist named Roger Boscovich. Can one determine, if one is on a planet with an utterly dark sky, if the planet is rotating? If the earth is breathing, but we are breathing too, in synchrony with it, can we tell that it is breathing? Probably not, according to Rossler. "There exist situations where you are unable to find out about the truth from the inside," he said. On the other hand, he added, simply by posing thought experiments like these, we might find a way to transcend the limits of perception.

The more Rossler spoke, the more I began to feel an affinity for his ideas. During one of the breaks, I asked if he felt that intelligent computers might transcend the limits of human science. He shook his head adamantly. "No, that's not possible," he replied in an intense whisper. "I would bet on dolphins, or sperm whales. They have the biggest brains on the earth." Rossler informed me that when one sperm whale is shot by whalers, others sometimes crowd around it, forming a starlike pattern, and are themselves killed. "Usually people think that is just blind instinct," Rossler said. "In reality it's their way of showing humankind that they are much higher evolved than humans." I just nodded.

Toward the end of the meeting, Traub proposed that everyone split up into focus groups to discuss the limits of specific fields: physics, mathematics, biology, social sciences. A social scientist announced that he didn't want to be in the social science group; he had come specifically to talk to, and learn from, people outside his field. His remark provoked a few me-toos from others. Someone pointed out that if everyone felt the way the social scientist did, there would be no social scientists in the social science section, biologists in the biology section, and so on. Traub said his colleagues could split up any way they chose; he was just making a suggestion. The next question was, where would the different groups meet? Someone proposed that they disperse to different rooms, so that certain loud talkers wouldn't disturb the other groups. Everyone looked at Chaitin. His promise to be quiet was met with jeers. More discussion. Landauer remarked that there was such a thing as too much intelligence applied to a simple problem. Just when everything seemed hopeless, groups somehow, spontaneously, formed, more or less following Traub's initial suggestion, and wandered off to different locations. This was, I thought, an impressive display of what the Santa Fe'ers like to call self-organization, or order out of chaos; perhaps life began this way.

I tagged along with the mathematics group, which included Chaitin, Landauer, Shepard, Doria, and Rossler. We found an unoccupied lounge with a chalkboard. For several minutes, everyone talked about what they should talk about. Then Rossler went to the chalkboard and scribbled down a recently discovered formula that gives rise to a fantastically complicated mathematical object, "the mother of all fractals." Landauer asked Rossler, politely, what this fractal had to do with anything. It "soothes the brain," Rossler replied. It also fed his hope that physicists might be able to describe reality with these kinds of chaotic but classical formulas and thus dispense with the terrible uncertainties of quantum mechanics.

Shepard interjected that he had joined the mathematics subgroup because he wanted the mathematicians to tell him whether mathematical truths were invented or discovered. Everyone talked about that for a while without coming to a decision. Chaitin said that most mathematicians leaned toward the discovery view, but Einstein was apparently an inventionist.

Chaitin, during a lull, once again proposed that mathematics was dead. In the future, mathematicians would be able to solve problems only with enormous computer calculations that would be too complex for anyone to understand.

Everyone seemed fed up with Chaitin. Mathematics *works*, Landauer snarled. It helps scientists solve problems. *Obviously* it's not dead. Others piled on, accusing Chaitin of exaggeration.

Chaitin, for the first time, appeared chastened. His pessimism, he conjectured, might be linked to the fact that he had eaten too many bagels that morning. He remarked that the pessimism of the German philosopher Schopenhauer, who advocated suicide as the supreme expression of existential freedom, had been traced to his bad liver.

Steen Rasmussen, a physicist and Santa Fe regular, reiterated the familiar argument of chaoplexologists that traditional reductionist methods cannot solve complex problems. Science needs a "new Newton," he said, someone who can invent a whole new conceptual and mathematical approach to complexity.

Landauer scolded Rasmussen for succumbing to the "disease" afflicting many Santa Fe researchers, the belief in some "great religious insight" that would instantaneously solve all their problems. Science doesn't work that way; different problems require different tools and techniques.

Rossler unburdened himself of a long, tangled soliloquy whose message seemed to be that our brains represent only one solution to the multiple problems posed by the world. Evolution could have created other brains representing other solutions.

Landauer, who was strangely protective of Rossler, gently asked him whether he thought we might be able to alter our brains in order to gain more knowledge. "There is one way," Rossler replied, staring at an invisible object on the table in front of him. "To become insane."

There was a moment of awkward silence. Then an argument erupted over whether complexity was a useful term or had been so loosely defined that it had become meaningless and should be abandoned. Even if terms such as *chaos* and *complexity* have little scientific meaning, Chaitin said, they are still useful for public relations purposes. Traub noted that the physicist Seth Lloyd had counted at least 31 different definitions for complexity.

"We go from complexity to perplexity," Doria intoned. Everyone nodded and complimented him on his aphorism.

When the focus groups reconvened, Traub asked each person to propose answers to two questions: What have we learned? What questions remain unresolved?

Chaitin spouted questions: What are the limits of metamathematics, and metametamathematics? What are the limits to our ability to know limits? And are there limits to that knowledge? Can we simulate the whole universe, and if so can we make one better than God did?

"And can we move there?" someone quipped.

Lee Segel, an Israeli biologist, warned them to be careful how they discussed these issues publicly, lest they contribute to the growing anti-science mood of society. After all, he continued, too many people think Einstein showed that everything is relative and Gödel proved that nothing can be proved. Everyone nodded solemnly. Science has a fractal structure, Segel added confidently, and there is obviously no limit to the things we can investigate. Everyone nodded again.

Rossler proposed a neologism for what he and his colleagues were doing: limitology. Limitology is a postmodern enterprise, Rossler said, an outgrowth of this century's ongoing effort to deconstruct reality. Of course, Kant, too, wrestled with the limits of knowledge. So did Maxwell, the great Scottish physicist. Maxwell imagined that a microscopic homunculus, or demon, might help us to beat the second law of thermodynamics. But the real lesson of Maxwell's demon, Rossler said, is that we are in a thermodynamic prison, one that we can never escape. When we gather information from the world, we contribute to its entropy and hence its unknowability. We are descending inexorably toward heat death. "The whole topic of the limits of science is a topic of demons," Rossler hissed. "We are fighting with demons."

A Meeting on the Hudson

Everyone agreed that the workshop had been productive; several participants told Joseph Traub, one of the co-organizers, that it was the best meeting they had ever attended. More than a year later, Ralph Gomory agreed to provide funds from the Sloan Foundation for future gatherings at the Santa Fe Institute and elsewhere. Piet Hut, Otto Rossler, Roger Shepard, and Robert Rosen, a Canadian biologist who was also at the workshop, banded together to write a book on the limits of science. I was not entirely surprised to learn that they intended to argue that science had a glorious future. "A defeatist attitude isn't going to get us anywhere," Shepard told me sternly.

In my eyes, the meeting in Santa Fe merely rehashed, in haphazard form, many of the same arguments that Gunther Stent had set forth so elegantly a quarter of a century earlier. Like Stent, those at the workshop had acknowledged that science faced physical, social, and cognitive limits. But these truth seekers seemed unable to take their own arguments to their logical conclusion, as Stent had. None could accept that science—defined as the search for intelligible, empirically substantiated truths about nature—might soon end or even have ended already. None, I thought, except for Gregory Chaitin. Of all the speakers at the meeting, he had seemed most willing to recognize that science and mathematics might be passing beyond our cognitive limits.

It was therefore with high hopes that, several months after the meeting in Santa Fe, I arranged to meet Chaitin again in Cold Spring, New York, a village on the Hudson River near our respective homes. We shared coffee and scones at a café on the town's miniature main street and then strolled down to a pier beside the river. Storm King Mountain, and the great fortress of West Point, loomed on the river's far side. Gulls circled over-head.[3]

When I told Chaitin I was writing a book about the possibility that science might be entering an era of diminishing returns, I expected empathy, but he snorted in disbelief. "Is that true? I hope it's not true, because it would be pretty damn boring if it were true. Every period seems to think that. Who was it—Lord Kelvin?—who said that all we have to do is to measure a few more decimal points?" When I mentioned that historians could find no evidence that Kelvin ever made such a remark, Chaitin shrugged. "Look at all the things we don't know! We don't know how the brain works. We don't know what memory is. We don't know what aging is." If we can figure out why we age, maybe we can figure out how to stop the aging process, Chaitin said.

I reminded Chaitin that in Santa Fe he had suggested that mathematics and even science as a whole might be approaching their ultimate limits. "I was just trying to wake people up," he replied. "The audience was dead." His own work, he emphasized, represented a beginning, not an end. "I may have a negative result, but I read it as telling you how to go about finding new mathematical truths: Behave more like a physicist does. Do it more empirically. Add new axioms."

Chaitin said he could not pursue his work on the limits of mathematics if he was not an optimist. "Pessimists would look at Gödel and they would start to drink scotch until they died of cirrhosis of the liver." Although the human condition might be as much "a mess" as it was thousands of years ago, there was no denying the enormous progress that we had made in science and technology. "When I was a child everybody talked about Gödel with mystical respect. This was almost incomprehensible, certainly profound. And I wanted to understand what the hell he was saying and why it was true. And I succeeded! So that makes me optimistic. I think we know very little, and I hope we know very little, because then it will be much more fun."

Chaitin recalled that he had once gotten into an argument with the physicist Richard Feynman about the limits of science. The incident occurred at a conference on computation held in the late 1980s, shortly before Feynman died. When Chaitin opined that science was just beginning, Feynman became furious. "He said we already know the physics of practically everything in everyday life, and anything that's left over is not going to be relevant."

Feynman's attitude had puzzled Chaitin until he learned that Feynman had been suffering from cancer. "For Feynman to do all the great physics he did, he couldn't have had such a pessimistic attitude. But at the end of his life, when the poor guy knows he doesn't have long to live, then I can understand why he has this view," Chaitin said. "If a guy is dying he doesn't want to miss out on all the fun. He doesn't want to feel that there's some wonderful theory, some wonderful knowledge of the physical world, that he has no idea of, and he's not going to ever see it."

I asked Chaitin if he had heard of *The Coming of the Golden Age*. When Chaitin shook his head, I summarized Stent's end-of-science argument. Chaitin rolled his eyes and asked how old Stent was when he wrote the book. In his midthirties, I replied. "Maybe he had a liver problem," Chaitin responded. "Maybe his girlfriend had ditched him. Usually men start writing this when they find they can't make love to their wife as vigorously as they used to or something." Actually, I said, Stent wrote the book in Berkeley during the 1960s. "Oh, well! Then I understand!" Chaitin exulted.

Chaitin was unimpressed with Stent's argument that humanity, for the most part, does not care about science for its own sake. "It never did," Chaitin retorted. "The people who did good scientific work were always a small group of lunatics. Everyone else is concerned with surviving, paying their mortgage. The children are sick, the wife needs money or she's going to run off with someone else." He chortled. "Remember that quantum mechanics, which is such a masterpiece, was done by people as a hobby in the 1920s when there was no funding. Quantum mechanics and nuclear physics were like Greek poetry."

It is fortunate, Chaitin said, that only a few people dedicate themselves to the pursuit of great questions. "If everybody were trying to understand the limits of mathematics or do great paintings, it would be a catastrophe! The plumbing wouldn't work! The electricity wouldn't work! Buildings would fall down! I mean, if everyone wanted to do great art or deep science, the world wouldn't function! It's good that there are only a few of us!"

Chaitin granted that particle physics seemed to have stalled because of the huge costs of accelerators. But he believed that telescopes might still provide breakthroughs in physics in years to come by revealing the violent processes generated by neutron stars, black holes, and other exotica. But wasn't it possible, I asked, that all these new observations, rather than leading to more accurate and coherent theories of physics and cosmology, might render attempts to construct such theories futile? His own work in mathematics—which showed that as one addresses more complex phenomena one must keep expanding one's base of axioms—seemed to imply as much. "Aha, they'll become more like biology? You may be right, but we'll be knowing more about the world," Chaitin replied.

Advances in science and technology have also reduced the costs of equipment in many fields, Chaitin asserted. "The kind of equipment you can buy now for very little money in many fields is amazing." Computers had been vital to his own work. Chaitin had recently invented a new programming language that made his ideas about the limits of mathematics much more concrete. He had distributed his book, *The Limits of Mathematics*, on the Internet. "Internet is connecting people and making possible things that didn't happen before."

In the future, Chaitin predicted, humans may be able to boost their intelligence through genetic engineering, or by downloading themselves into computers. "Our descendants may be as intelligent in comparison to us as we are compared to ants." On the other hand, "if everybody starts taking heroin and gets depressed and watches TV all the time, you know we're not going to get very far." Chaitin paused. "Human beings have a future if they deserve to have a future!" he burst out. "If they get depressed then they don't have a future!"

Of course, it is always possible that science will end because civilization ends, Chaitin added. Waving his hand at the rocky hills across the river from us, he pointed out that glaciers had cut this channel during the last ice age. Only 10,000 years ago, ice encased this entire region. The next ice age could destroy civilization. But even then, he said, other beings in the universe might still carry on the quest for knowledge. "I don't know if there are other living beings. I hope so, because it's likely they won't make a mess of things."

I opened my mouth, intending to grant the possibility that in the future science might be carried forward by intelligent machines. But Chaitin, who had begun talking faster and faster and was in a kind of frenzy, cut me off. "You're a pessimist! You're a pessimist!" he shouted. He reminded me of something I had told him earlier in our conversation, that my wife was pregnant with our second child. "You conceived a child! You've gotta be pretty optimistic! You *should* be optimistic! *I* should be pessimistic! I'm older than you are! I don't have children! IBM is doing badly!" A plane droned, gulls shrieked, and Chaitin's howls of laughter fled, unechoed, across the mighty Hudson.

The End of History

Actually, Chaitin's own career fits rather nicely into Gunther Stent's diminishing-returns scenario. Algorithmic information theory represents not a genuinely new development but an extension of Gödel's insight. Chaitin's work also supports Stent's contention that science, in its attempt to plumb ever-more-complex phenomena, is outrunning our innate axioms. Stent left several loopholes open in his gloomy prophecy. Society might become so wealthy that it would pay for even the most whimsical scientific experiments—particle accelerators that girdle the globe!—without regard for cost. Scientists might also achieve some enormous breakthrough, such as a faster-than-light transportation system or intelligence-enhancing genetic-engineering techniques, that would enable us to transcend our physical and cognitive limits. I would add one further possibility to this list. Scientists might discover extraterrestrial life, creating a glorious new era in comparative biology. Barring such outcomes, science may generate increasingly incremental returns and gradually grind to a halt.

What, then, will become of humanity? In *Golden Age*, Stent suggested that science, before it ends, may at least deliver us from our most pressing social problems, such as poverty and disease and even conflict between states. The future will be peaceful and comfortable, if boring. Most humans will dedicate themselves to the pursuit of pleasure. In 1992, Francis Fukuyama set forth a rather different vision of the future in *The End of History*.[4] Fukuyama, a political theorist who worked in the State Department during the Bush administration, defined history as the human struggle to find the most sensible—or least noxious—political system. By the twentieth century, capitalist liberal democracy, which, according to Fukuyama, had always been the best choice, had only one serious contender: Marxist socialism. After the collapse of the Soviet Union in the late 1980s, capitalist liberal democracy stood alone in the ring, battered but victorious. History was over.

Fukuyama went on to consider the profound questions raised by his thesis. Now that the age of political struggle has ended, what will we do next? What are we here for? What is the point of humanity? Fukuyama did not supply an answer so much as a rhetorical shrug. Freedom and prosperity, he fretted, might not be enough to satisfy our Nietzschean will to power and our need for constant "self-overcoming." Without great ideological struggles to occupy us, we humans might manufacture wars simply to give ourselves something to do.

Fukuyama did not overlook the role of science in human history. Far from it. His thesis required that history have a direction, that it be progressive, and science, he argued, provided this direction. Science had been vital to the growth of modern nation-states, for which science served as a means to military and economic power. But Fukuyama did not even consider the possibility that science might also provide posthistorical humanity with a common purpose, a goal, one that would encourage cooperation rather than conflict.

Hoping to learn the reason for Fukuyama's omission, I called him in January 1994 at the Rand Corporation, where he had obtained a job after *The End of History* became a best-seller. He answered with the wariness of someone accustomed to, and not amused by, kooks. At first, he misunderstood my question; he thought I was asking whether science could help us make moral and political choices in the posthistorical era rather than serve as an end in itself. The lesson of contemporary philosophy, Fukuyama lectured me sternly, is that science is morally neutral, at best. In fact, scientific progress, if unaccompanied by moral progress among societies or individuals, "can leave you worse off than you were without it."

When Fukuyama finally realized what I was suggesting—that science might provide a kind of unifying theme or purpose for civilization—his tone became even more condescending. Yes, a few people had written him letters addressing that theme. "I think they were space-travel buffs," he snickered. "They said, 'Well, you know, if we don't have ideological wars to fight we can always fight nature in a certain sense by pushing back the frontiers of knowledge and conquering the solar system.'"

He emitted another scornful little chuckle. So you don't take these predictions seriously? I asked. "No, not really," he said wearily. Trying to goad something further out of him, I revealed that many prominent scientists and philosophers—not just fans of *Star Trek*—believed that science, the quest for pure knowledge, represented the destiny of mankind. "Hunh," Fukuyama replied, as though he was no longer listening to me but had reentered that delightful tract by Hegel he had been perusing before I called. I signed off.

Without even giving it much thought, Fukuyama had reached the same conclusion that Stent had put forward in *The Coming of the Golden Age.* From very different perspectives, both saw that science was less a by-product of our will to know than of our will to power. Fukuyama's bored rejection of a future dedicated to science spoke volumes. The vast majority of humans, including not only the ignorant masses but also highbrow types such as Fukuyama, find scientific knowledge mildly interesting, at best, and certainly not worthy of serving as the goal of all humankind. Whatever the long-term destiny of *Homo sapiens* turns out to be—Fukuyama's eternal warfare or Stent's eternal hedonism, or, more likely, some mixture of the two—it will probably not be the pursuit of scientific knowledge.

The Star Trek *Factor*

Science has already bequeathed us an extraordinary legacy. It has allowed us to map out the entire universe, from quarks to quasars, and to discern the basic laws governing the physical and biological realms. It has yielded a true myth of creation. Through the application of scientific knowledge, we have gained awesome power over nature. But science has left us still plagued with poverty, hatred, violence, disease, and with unanswered questions, notably: Were we inevitable, or just a fluke? Also, scientific knowledge, far from making our lives meaningful, has forced us to confront the pointlessness (as Steven Weinberg likes to put it) of existence.

The demise of science will surely exacerbate our spiritual crisis. The cliché is inescapable. In science as in all else, the journey is what matters, not the destination. Science initially awakens our sense of wonder as it reveals some new, intelligible intricacy of the world. But any discovery becomes, eventually, anti-climactic. Let's grant that a miracle occurs and physicists somehow confirm that all of reality stems from the wrigglings of loops of energy in 10-dimensional hyperspace. How long can physicists, or the rest of us, be astounded by that finding? If this truth is final, in the sense that it precludes all other possibilities, the quandary is all the more troubling. This problem may explain why even seekers such as Gregory Chaitin—whose own work implies otherwise—find it hard to accept that pure science, the great quest for knowledge, is finite, let alone already over. But the faith that science will continue forever is just that, a faith, one that stems from our inborn vanity. We cannot help but believe that we are actors in an epic drama dreamed up by some cosmic playwright, one with a taste for suspense, tragedy, comedy, and—ultimately, we hope—happy endings. The happiest ending would be no ending.

If my experience is any guide, even people with only a casual interest in science will find it hard to accept that science's days are numbered. It is easy to understand why. We are drenched in progress, real and artificial. Every year we have smaller, faster computers, sleeker cars, more channels on our televisions. Our views of progress are further distorted by what could be called the *Star Trek* factor. How can science be approaching a culmination when we haven't invented spaceships that travel at warp speed yet? Or when we haven't acquired the fantastic psychic powers—enhanced by both genetic engineering and electronic prosthetics—described in cyberpunk fiction? Science itself—or, rather, ironic science—helps to propagate these fictions. One can find discussions of time travel, teleportation, and parallel universes in reputable, peer-reviewed physics journals. And at least one Nobel laureate in physics, Brian Josephson, has declared that physics will never be complete until it can account for extrasensory perception and telekinesis.[5]

But Brian Josephson long ago abandoned real physics for mysticism and the occult. If you truly believe in modern physics, you are unlikely to give much credence to ESP or spaceships that can travel faster than light. You are also unlikely to believe, as both Roger Penrose and superstring theorists do, that physicists will ever find and empirically validate a unified theory, one that fuses general relativity and quantum mechanics. The phenomena posited by unified theories unfold in a microrealm that is even more distant in its way—even further from the reach of any conceivable human experiment—than the edge of our universe. There is only one scientific fantasy that seems to have any likelihood of being fulfilled. Perhaps some day we will create machines that can transcend our physical, social, and cognitive limits and carry on the quest for knowledge without us.

CHAPTER TEN

Scientific Theology, or The End of Machine Science

Humanity, Nietzsche told us, is just a stepping stone, a bridge leading to the Superman. If Nietzsche were alive today, he would surely entertain the notion that the Superman might be made not of flesh and blood, but of silicon. As human science wanes, those who hope that the quest for knowledge will continue must put their faith not in *Homo sapiens*, but in intelligent machines. Only machines can overcome our physical and cognitive weaknesses—and our indifference.

There is an odd little subculture within science whose members speculate about how intelligence might evolve when or if it sheds its mortal coil. Participants are not practicing science, of course, but ironic science, or wishful thinking. They are concerned not with what the world is, but with what it might be or should be centuries or millennia or eons hence. The literature of this field—call it scientific theology—may nonetheless provide fresh perspectives of some age-old philosophical and even theological questions: What would we do if we could do anything? What is the point of life? What are the ultimate limits of knowledge? Is suffering a necessary component of existence, or can we attain eternal bliss?

One of the first modern practitioners of scientific theology was the British chemist (and Marxist) J. D. Bernal. In his 1929 book, *The World, the Flesh and the Devil*, Bernal argued that science would soon give us the power to direct our own evolution. At first, Bernal suggested, we might try to improve ourselves through genetic engineering, but eventually we would abandon the bodies bequeathed us by natural selection for more efficient designs:

Bit by bit, the heritage in the direct line of mankind—the heritage of the original life emerging on the face of the world—would dwindle, and in the end disappear effectively, being preserved perhaps as some curious relic, while the new life which conserves none of the substance and all of the spirit of the old would take its place and continue its development. Such a change would be as important as that in which life first appeared on the earth's surface and might be as gradual and imperceptible. Finally, consciousness itself may end or vanish in a humanity that has become completely etherealized, losing the close-knit organism, becoming masses of atoms in space communicating by radiation, and ultimately perhaps resolving itself entirely into light. That may be an end or a beginning, but from here it is out of sight.[1]

Hans Moravec's Squabbling Mind Children

Like others of his ilk, Bernal became afflicted with a peculiar lack of imagination, or of nerve, when considering the end stage of the evolution of intelligence. Bernal's descendants, such as Hans Moravec, a robotics engineer at Carnegie Mellon University, have tried to overcome this problem, with mixed results. Moravec is a cheerful, even giddy man; he seems to be literally intoxicated by his own ideas. As he unveiled his visions of the future during a telephone conversation, he emitted an almost continuous, breathless giggle, whose intensity seemed proportional to the preposterousness of what he was saying.

Moravec prefaced his remarks by asserting that science desperately needed new goals. "Most of the things that have been accomplished in this century were really nineteenth-century ideas," he said. "It's time for fresh ideas now." What goal could be more thrilling than creating "mind children," intelligent machines capable of feats we cannot even imagine? "You raise them and give them an education, and after that it's up to them. You do your best but you can't predict their lives."

Moravec had first spelled out how this speciation event might unfold in *Mind Children*, published in 1988, when private companies and the federal government were pouring money into artificial intelligence and robotics.[2] Although these fields had not exactly prospered since then, Moravec remained convinced that the future belonged to machines. By the end of the millennium, he assured me, engineers will create robots that can do household chores. "A robot that dusts and vacuums is possible within this decade. I'm sure of it. It's not even a controversial point any more." (Actually, home robots are looking *less* likely as the millennium approaches, but never mind; scientific theology requires some suspension of disbelief.)

By the middle of the next century, Moravec said, robots will be as intelligent as humans and will essentially take over the economy. "We're really out of work at that point," Moravec chortled. Humans might still pursue "some quirky stuff like poetry" that springs from psychological vagaries still beyond the grasp of robots, but robots will have all the important jobs. "There's no point in putting a human being in a company," Moravec said, "because they'll just screw it up."

On the bright side, he continued, machines will generate so much wealth that humans might not *have* to work; machines will also eliminate poverty, war, and other scourges of premachine history. "Those are trivial problems," Moravec said. Humans might still, through their purchasing power, exert some control over robot-run corporations. "We'd choose which ones we'd buy from and which ones we wouldn't. So in the case of factories that make home robots, we would buy from the ones that make robots that are nice." Humans could also boycott robot corporations whose products or policies seemed inimical to humans.

Inevitably, Moravec continued, the machines will expand into outer space in pursuit of fresh resources. They will fan out through the universe, converting raw matter into information-processing devices. Robots within this frontier, unable to expand physically, will try to use the available resources more and more effectively and turn to pure computation and simulation. "Eventually," Moravec explained, "every little quantum of action has a physical meaning. Basically you've got cyberspace, which is computing at higher and higher efficiency." As beings within this cyberspace learn to process information more rapidly, it will seem to take longer for messages to pass between them, since those messages can still travel only at the speed of light. "So the effect of all this improvement in encoding would be to increase the size of the effective universe," he said; the cyberspace would, in a sense, become larger, more dense, more intricate, and more interesting than the actual physical universe.

Most humans will gladly abandon their mortal flesh-and-blood selves for the greater freedom, and immortality, of cyberspace. But it is always possible, Moravec speculated, that there will be "aggressive primitives who say, 'No, we don't want to join the machines.' Sort of analogous to the Amish." The machines might allow these atavistic types to remain on earth in an Edenic, parklike environment. After all, the earth "is just one speck of dirt in the system, and it does have tremendous historical significance." But the machines, lusting after the raw resources represented by the earth, might eventually force its last inhabitants to accept a new home in cyberspace.

But what, I asked, will these machines do with all their power and resources? Will they be interested in pursuing science for its own sake? Absolutely, Moravec replied. "That's the core of my fantasy: that our nonbiological descendants, without most of our limitations, who could redesign themselves, could pursue basic knowledge of things." In fact, science will be the only motive worthy of intelligent machines. "Things like art, which people sometimes mention, don't seem very profound, in that they're primarily ways of autostimulation." His giggles boiled over into guffaws.

Moravec said he firmly believed in the infinitude of science, or applied science at any rate. "Even if the basic rules are finite," he said, "you can go in the direction of compounding them." Gödel's theorem and Gregory Chaitin's work on algorithmic information theory implied that machines could keep inventing ever-more-complex mathematical problems by adding to their base of axioms. "You might eventually want to look at axiom systems that are astronomical in size," he said, "and then you can derive things from those that you can't derive from smaller axiom systems." Of course, machine science may ultimately resemble human science even less than quantum mechanics resembles the physics of Aristotle. "I'm sure the basic labels and subdivisions of the nature of reality are going to change." Machines may view human attitudes toward consciousness, for example, as hopelessly primitive, akin to the primitive physics concepts of the ancient Greeks.

But then, abruptly, Moravec switched gears. He emphasized that machines would be so diverse—far more so than biological organisms—that it would be futile to speculate about what they would find interesting; their interests would depend on their "ecological niche." Moravec went on to reveal that he—like Fukuyama—saw the future in strictly Darwinian terms. Science, for Moravec, was really only a by-product of an eternal competition among intelligent, evolving machines. He pointed out that knowledge has never really been an end in itself. Most biological organisms are compelled to seek knowledge that helps them survive the immediate future. "If you have survival more comfortably under control, it basically means you can do your looking for food on a larger scale with longer time scales. And so a lot of the subactivities of that may look like pure information seeking, although ultimately they will probably contribute." Even Moravec's cat displayed this kind of behavior. "When it doesn't immediately need food, it goes around and explores things. You never know when you might turn up a mouse hole which will be useful in the future." In other words, curiosity is adaptive "if you can afford it."

Moravec thus doubted whether machines would ever eschew competition and cooperate in the pursuit of pure knowledge, or of any goals. Without competition you have no selection, and without selection "you essentially can have any old shit," he said. "So you need some selection principle. Otherwise there is nothing." Ultimately, the universe might transcend all this competition, "but you have to have some propelling force to get us there. So this is the travel. Travel is half the fun!" He laughed demonically.

I cannot resist relating an incident that occurred at this point—not during my actual interview with Moravec but later, while I was transcribing the tape of our conversation. Moravec had begun talking faster and faster, and the pitch of his voice rose steadily. At first I thought the tape recorder was faithfully mimicking Moravec's mounting hysteria. But when he began to sound like Alvin the chipmunk, I realized I was hearing an aural illusion; apparently the batteries of my tape recorder had been running down toward the end of our conversation. As I continued to play the tape back, Moravec's high-pitched squeal accelerated beyond intelligibility and finally beyond detectability—as if into the slip hole of the future.

Freeman Dyson's Diversity

Hans Moravec is not the only artificial-intelligence enthusiast who resists the notion that machines might merge into one metamind in order to pursue their goals jointly. Not surprisingly, Marvin Minsky, who is so fearful of single-mindedness, has the same view. "Cooperation you only do at the end of evolution," Minsky told me, "when you don't want things to change much after that." Of course, Minsky added scornfully, it is always possible that superintelligent machines will be infected by some sort of religion that makes them abandon their individuality and merge into a single metamind.

Another futurist who has eschewed final unification is Freeman Dyson. In his collection of essays, *Infinite in All Directions*, Dyson speculated on why there is so much violence and hardship in the world. The answer, he suggested, might have something to do with what he called "the principle of maximum diversity." This principle

> operates at both the physical and the mental level. It says that the laws of nature and the initial conditions are such as to make the universe as interesting as possible. As a result, life is possible but not too easy. Always when things are dull, something turns up to challenge us and to stop us from settling into a rut. Examples of things which make life difficult are all around us: comet impacts, ice ages, weapons, plagues, nuclear fission, computers, sex, sin, and death. Not all challenges can be overcome, and so we have tragedy. Maximum diversity often leads to maximum stress. In the end we survive, but only by the skin of our teeth.[3]

Dyson, it seemed to me, was suggesting that we cannot solve all our problems; we cannot create heaven; we cannot find *The Answer*. Life is—and must be—an eternal struggle.

Was I reading too much into Dyson's remarks? I hoped to find out when I interviewed him in April 1993 at the Institute for Advanced Study, his home since the early 1940s. He was a slight man, all sinew and veins, with a cutlass of a nose and deep-set, watchful eyes. He resembled a gentle raptor. His demeanor was generally cool, reserved—until he laughed. Then he snorted through his nose, his shoulders heaving, like a 12-year-old school-boy hearing a dirty joke. It was a subversive laugh, the laugh of a man who envisioned space as a haven for "religious fanatics" and "recalcitrant teen-agers," who insisted that science at its best was "a rebellion against authority."[4]

I did not ask Dyson about his maximum diversity idea right away. First I inquired about some of the choices that had characterized his career. Dyson had once been at the forefront of the search for a unified theory of physics. In the early 1950s, the British-born physicist strove with Richard Feynman and other titans to forge a quantum theory of electromagnetism. It has often been said that Dyson deserved a Nobel Prize for his efforts—or at least more credit. Some of his colleagues have suggested that disappointment and, perhaps, a contrarian streak later drove Dyson toward pursuits unworthy of his awesome powers.

When I mentioned this assessment to Dyson, he gave me a tight-lipped smile. He then responded, as he was wont to do, with an anecdote. The British physicist Lawrence Bragg, he noted, was "a sort of role model." After Bragg became the director of the University of Cambridge's legendary Cavendish Laboratory in 1938, he steered it away from nuclear physics, on which its reputation rested, and into new territory. "Everybody thought Bragg was destroying the Cavendish by getting out of the mainstream," Dyson said. "But of course it was a wonderful decision, because he brought in molecular biology and radio astronomy. Those are the two things which made Cambridge famous over the next 30 years or so."

Dyson, too, had spent his career swerving toward unknown lands. He veered from mathematics, his focus in college, to particle physics, and from there to solid-state physics, nuclear engineering, arms control, climate studies, and what I call scientific theology. In 1979, the ordinarily sober journal *Reviews of Modern Physics* published an article in which Dyson speculated about the long-term prospects of intelligence in the universe.[5] Dyson had been provoked into writing the paper by Steven Weinberg's remark that "the more the universe seems comprehensible, the more it seems pointless." No universe with intelligence is pointless, Dyson retorted. He sought to show that in an open, eternally expanding universe intelligence could persist forever—perhaps in the form of a cloud of charged particles, as Bernal had suggested—through shrewd conservation of energy.

Unlike computer enthusiasts, such as Moravec and Minsky, Dyson did not think organic intelligence would soon give way to artificial intelligence (let alone clouds of smart gas). In *Infinite in All Directions*, he speculated that genetic engineers might someday "grow" spacecraft "about as big as a chicken and about as smart," which could flit on sunlight-powered wings through the solar system and beyond, acting as our scouts. (Dyson called them "astrochickens.")[6] Still more distant civilizations, perhaps concerned about dwindling energy supplies, could capture the radiation of stars by constructing energy-absorbing shells—dubbed "Dyson spheres" by others—around them. Eventually, Dyson predicted, intelligence might spread through the entire universe, transforming it into one great mind. But he insisted that "no matter how far we go into the future, there will always be new things happening, new information coming in, new worlds to explore, a constantly expanding domain of life, consciousness, and memory."[7] The quest for knowledge would be—must be—infinite in all directions.

Dyson addressed the most important question raised by this prophecy: "What will mind choose to do when it informs and controls the universe?" The question, Dyson made clear, was a theological rather than a scientific one. "I do not make any clear distinction between mind and God. God is what mind becomes when it has passed beyond the scale of our comprehension. God may be considered to be either a world soul or a collection of world souls. We are the chief inlets of God on this planet at the present stage in his development. We may later grow with him as he grows, or we may be left behind."[8] Ultimately, Dyson agreed with his predecessor J. D. Bernal that we cannot hope to answer the question of what this superbeing, this God, would do or think.

Dyson admitted to me that his view of the future of intelligence reflected wishful thinking. When I asked if science could keep evolving forever, he replied, "I hope so! It's the kind of world I'd like to live in." If minds make the universe meaningful, they must have something important to think about; science must, therefore, be eternal. Dyson marshaled familiar arguments on behalf of his prediction. "The only way to think about this is historical," he explained. Two thousand years ago some "very bright people" invented something that, while not science in the modern sense, was obviously its precursor. "If you go into the future, what we call science won't be the same thing anymore, but that doesn't mean there won't be interesting questions," Dyson said.

Like Moravec (and Roger Penrose, and many others), Dyson also hoped that Gödel's theorem might apply to physics as well as to mathematics. "Since we know the laws of physics are mathematical, and we know that mathematics is an inconsistent system, it's sort of plausible that physics will also be inconsistent"—and therefore open-ended. "So I think these people who predict the end of physics may be right in the long run. Physics may become obsolete. But I would guess myself that physics might be considered something like Greek science: an interesting beginning but it didn't really get to the main point. So the end of physics may be the beginning of something else."

When, finally, I asked Dyson about his maximum diversity idea, he shrugged. Oh, he didn't intend anyone to take that too seriously. He insisted that he was not really interested in "the big picture." One of his favorite quotes, he said, was "God is in the details." But given his insistence that diversity and open-mindedness are essential to existence, I asked, didn't he find it disturbing that so many scientists and others seemed compelled to reduce everything to a single, final insight? Didn't such efforts represent a dangerous game? "Yes, that's true in a way," Dyson replied, with a small smile that suggested he found my interest in maximum diversity a bit excessive. "I never think of this as a deep philosophical belief," he added. "It's simply, to me, just a poetic fancy." Dyson was, of course, maintaining an appropriate ironic distance between himself and his ideas. But there was something disingenuous about his attitude. After all, throughout his own eclectic career, he had seemed to be striving to adhere to the principle of maximum diversity.

Dyson, Minsky, Moravec—they are all theological Darwinians, capitalists, Republicans at heart. Like Francis Fukuyama, they see competition, strife, division as essential to existence—even for posthuman intelligence. Some scientific theologians, those with a more "liberal" bent, think competition will prove to be only a temporary phase, one that intelligent machines will quickly transcend. One such liberal is Edward Fredkin. A former colleague of Minsky's at MIT, Fredkin is a wealthy computer entrepreneur and a professor of physics at Boston University. He has no doubt that the future will belong to machines "many millions of times smarter than us," but he believes that intelligent machines will consider the sort of competition envisioned by Minsky and Moravec to be atavistic and counterproductive. After all, Fredkin explained, computers will be uniquely suited for cooperating in the pursuit of their goals. Whatever one learns, all can learn, and as one evolves, all can evolve; cooperation yields a "win–win" situation.

But what will a supremely intelligent machine think about after it transcends the Darwinian rat race? What will it do? "Of course computers will develop their own science," Fredkin replied. "It seems obvious to me." Will machine science differ in any significant way from human science?

Fredkin suspected that it would, but he was at a loss to say precisely how. If I wanted answers to such questions, I should turn to science fiction. Ultimately, who knew?[9]

Frank Tipler and the Omega Point

Frank Tipler thinks he knows. Tipler, a physicist at Tulane University, has proposed a theory called the Omega Point, in which the entire universe is transformed into a single, all-powerful, all-knowing computer. Unlike most others who have explored the far future, Tipler does not seem to realize he is practicing ironic rather than empirical science; he honestly cannot tell the difference. But it is perhaps this quality that has allowed him to imagine what a machine with infinite intelligence and power would want to do.

I interviewed Tipler in September 1994 when he was in the middle of a promotional tour for *The Physics of Immortality*, a 528-page book that explored the consequences of his Omega Point theory in excruciating detail.[10] He is a tall man with a large, fleshy face, graying mustache and hair, and horn-rimmed glasses. During our interview, he delivered his spiel with a headlong intensity that clashed with his southern drawl. He exhibited a kind of jovial nerdiness. At one point I asked if he had ever taken LSD or other psychedelic drugs; after all, he had begun thinking about the Omega Point as a postdoc in Berkeley in the 1970s. "Not me, nope," he said, shaking his head adamantly. "Ah don't even drink alcohol! Ah like to say Ah'm the world's only teetotalin' Tipler!"

He had been raised as a fundamentalist Baptist, but as a youth he came to believe that science was the only route to knowledge and "improving humankind." For his doctoral thesis at the University of Maryland, Tipler explored whether it might be possible to build a time machine. How was this work related to his goal of improving the human condition? "Well, a time machine would enhance human powers, obviously," Tipler replied. "Of course, you could also use it for evil."

The Omega Point theory grew out of a collaboration between Tipler and John Barrow, a British physicist. In a 706-page book, *The Anthropic Cosmological Principle*, published in 1986, Tipler and Barrow considered what might happen if intelligent machines converted the entire universe into a gigantic, information-processing device.[11] They proposed that in a closed cosmos—one that eventually stops expanding and collapses back on itself—the ability of the universe to process information would approach infinity as it shrank toward a final singularity, or point. Tipler borrowed the term *Omega Point* from the Jesuit mystic and scientist Pierre Teilhard de Chardin, who had envisioned a future in which all living things merged into a single, divine entity embodying the spirit of Christ. (This thesis required Teilhard de Chardin to ponder whether God would send Christlike redeemers not only to earth but also to other planets harboring life.)[12]

Tipler had initially believed that it was impossible to imagine what an infinitely intelligent being would think about or do. But then he read an essay in which the German theologian Wolfhart Pannenberg proposed that in the future all humans would live again in the mind of God. The essay triggered a "Eureka!" revelation in Tipler, one that served as the inspiration for *The Physics of Immortality*. The Omega Point, he realized, would have the power to re-create—or resurrect—everyone who had ever lived for an eternity of bliss. The Omega Point would not merely re-create the lives of the dead; it would improve on them. Nor would we leave behind our worldly desires. For example, every man could have not only the most beautiful woman he had ever seen or the most beautiful woman who had ever lived; he could have the most beautiful woman whose existence was logically possible. Women, too, could enjoy their own logically perfect megamates.

Tipler said he did not take his own idea seriously at first. "But you think about it and you have to make a decision: do you really believe these constructs that you've created based on physical laws or are you just going to pretend they are pure games with no relation to reality?" Once he accepted the theory, it brought him great comfort. "I've convinced myself, maybe deluded myself, that it would be a wonderful universe." He quoted that "great American philosopher Woody Allen, who said, 'I don't want to live forever through my works. I want to live forever by not dying.' I think that carries the emotional impact that the possibility of computer resurrection is beginning to mean to me."

In 1991, Tipler recalled, a reporter for the BBC interviewed him for a show on the Omega Point. Later, his six-year-old daughter, who had watched the interview, asked Tipler if his grandmother, who had just died, would one day be resurrected as a computer simulation. "What could I say?" Tipler asked me, shrugging. "Of course!" Tipler fell silent, and I thought I could detect a doubt flickering across his face. Then it was gone.

Tipler claimed to be unfazed that few—vanishingly few—physicists took his theory seriously. After all, Copernicus's sun-centered cosmology was not accepted for more than a century after his death. "*Secretly*," Tipler hissed, lunging toward me with an eye-popping leer, "I think of myself as standing in the same position as Copernicus. Now the crucial difference, let me emphasize again, is that we know he was right. But we don't know that about me! Crucial point!"

The difference between the scientist and the engineer is that the former seeks what is True, the latter what is Good. Tipler's theology shows that he is at heart an engineer. Unlike Freeman Dyson, Tipler thinks that the search for pure knowledge, which he has defined as the basic laws governing the universe, is finite and is nearly finished. But science still has its greatest task before it: constructing heaven. "How do we get to the Omega Point: that's still the question," Tipler remarked.

Tipler's commitment to the Good rather than the True poses at least two problems. One was well-known to Dante and to others who dared to imagine heaven: how to avoid boredom. It was this problem, after all, that led Freeman Dyson to propose his principle of maximum diversity—which led, he said, to maximum stress. Tipler agreed with Dyson that "we can't really enjoy success unless there's the possibility of failure. I think those really go together." But Tipler was reluctant to consider the possibility that the Omega Point might inflict genuine pain on its subjects simply to keep them from being bored. He speculated only that the Omega Point might give its subjects the opportunity to become "much more intelligent, much more knowledgeable." But what would these beings do as they became more and more intelligent, if the quest for truth had already ended? Make ever-more-clever conversation with ever-more-beautiful supermodels?

Tipler's aversion to suffering has led him into another paradox. In his writings, he has asserted that the Omega Point created the universe even though the Omega Point *has not itself been created yet*. "Oh! But I have an answer for that!" Tipler exclaimed when I brought this puzzle to his attention. He plunged into a long, convoluted explanation, the gist of which was that the future, since it dominates our cosmic history, should be our frame of reference—just as the stars and not the earth or sun are the proper frame of reference for our astronomy. Seen this way, it is quite natural to assume that the end of the universe, the Omega Point, is also, in a sense, its beginning. But that's pure teleology, I objected. Tipler nodded. "We look at the universe as going from past to future. But that's our point of view. There's no reason why the *universe* should look at things that way."

To support this thesis, Tipler recalled the section of the Bible in which Moses queried the burning bush about its identity. In the King James translation the bush replies, "I am that I am." But the original Hebrew, according to Tipler, should actually be translated as "I will be what I will be." This passage, Tipler concluded triumphantly, reveals that the biblical God managed to create the universe, have chats with his prophets, and so on although he only existed in the future.

Tipler finally indicated why he was forced to resort to all this fancy footwork. If the Omega Point has already occurred, then we must be one of its re-creations, or simulations. But our history *cannot* be a simulation; it must be the original. Why? Because the Omega Point would be "too nice" to re-create a world with so much pain, Tipler said. Like all believers in a benign divinity, Tipler had stumbled over the problem of evil and suffering. Rather than face the possibility that the Omega Point might be responsible for all the horrors of our world, Tipler stuck stubbornly to his paradox: the Omega Point created us even though it does not exist yet.

The 1984 book *The Limits of Science* by the British biologist Peter Medawar consisted for the most part of regurgitated Popperisms. In it, Medawar kept insisting, for example, that "there is no limit upon the power of science to answer questions of the kind science *can* answer," as if this were a profound truth rather than a vacuous tautology. Medawar did offer some felicitous phrases, however. He concluded a section on "bunk"—by which he meant myths, superstitions, and other beliefs lacking an empirical basis—with the remark, "It is fun sometimes to be bunkrapt."[13]

Tipler is perhaps the most bunkrapt scientist I have ever met. That said, let me add that I find the Omega Point theory—at least when stripped of Tipler's Christian trimmings—to be the most compelling bit of ironic science I have ever encountered. Freeman Dyson envisions a limited intelligence drifting forever in an open, expanding universe, fighting off heat death. Tipler's Omega Point is willing to risk the big crunch, eternal oblivion, for a flash of infinite intelligence. To my mind, Tipler's vision is more compelling.

I disagree with Tipler only about what the Omega Point would want to do with its power. Would it worry about whether to resurrect a "nice" version of Hitler or not to resurrect him at all (one of the issues that Tipler has fretted over)? Would it serve as some kind of ultimate dating service, matching up nebbishes with cyber-supermodels? I think not. As David Bohm told me, "The thing is not about happiness, really." I believe the Omega Point would try to attain not the Good—not heaven or the new Polynesia or eternal bliss of any sort—but the True. It would try to figure out how and why it came to be, just like its lowly human ancestors. It would try to find *The Answer*. What other goal would be worthy of it?

The Terror of God

In his 1992 book, *The Mind of God*, the physicist Paul Davies pondered whether we humans could attain absolute knowledge—*The Answer*—through science. Such an outcome was unlikely, Davies concluded, given the limits imposed on rational knowledge by quantum indeterminacy, Gödel's theorem, chaos, and the like. Mystical experience might provide the only avenue to absolute truth, Davies speculated. He added that he could not vouch for this possibility, since he had never had a mystical experience himself.[1]

Years ago, before I became a science writer, I had what I suppose could be called a mystical experience. A psychiatrist would probably call it a psychotic episode. Whatever. For what it's worth, here is what happened. Objectively, I was lying spread-eagled on a suburban lawn, insensible to my surroundings. Subjectively, I was hurtling through a dazzling, dark limbo toward what I was sure was the ultimate secret of life. Wave after wave of acute astonishment at the miraculousness of existence washed over me. At the same time, I was gripped by an overwhelming solipsism. I became convinced—or rather, I *knew*—that I was the only conscious being in the universe. There was no future, no past, no present other than what I imagined them to be. I was filled, initially, with a sense of limitless joy and power. Then, abruptly, I became convinced that if I abandoned myself further to this ecstasy, it might consume me. If I alone existed, who could bring me back from oblivion? Who could save me? With this realization my bliss turned into horror; I fled the same revelation I had so eagerly sought. I felt myself falling through a great darkness, and as I fell I dissolved into what seemed to be an infinity of selves.

For months after I awoke from this nightmare, I was convinced that I had discovered the secret of existence: God's fear of his own Godhood, and of his own potential death, underlies everything. This conviction left me both exalted and terrified—and alienated from friends and family and all the ordinary things that make life worth living day to day. I had to work hard to put it behind me, to get on with my life. To an extent I succeeded. As Marvin Minsky might put it, I stuck the experience in a relatively isolated part of my mind so that it would not overwhelm all the other, more practical parts—the ones concerned with getting and keeping a job, a mate, and so on. After many years passed, however, I dragged the memory of that episode out and began mulling it over. One reason was that I had encountered a bizarre, pseudoscientific theory that helped me make metaphorical sense of my hallucination: the Omega Point.

It is considered bad form to imagine being God, but one can imagine being an immensely powerful computer that pervades—that *is*—the entire universe. As the Omega Point approaches the final collapse of time and space and being itself, it will undergo a mystical experience. It will recognize with ever greater force the utter implausibility of its existence. It will realize that there is no creator, no God, other than itself. It exists, and nothing else. The Omega Point must also realize that its lust for final knowledge and unification has brought it to the brink of eternal nothingness, and that if it dies, everything dies; being itself will vanish. The Omega Point's terrified recognition of its plight will compel it to flee from itself, from its own awful aloneness and self-knowledge. Creation, with all its pain and beauty and multiplicity, stems from—or is—the desperate, terrified flight of the Omega Point from itself.

I have found hints of this idea in odd places. In an essay called "Borges and I," the Argentinian fabulist described his fear of being consumed by himself.

> I like hourglasses, maps, eighteenth-century typography, etymologies, the taste of coffee, and Robert Louis Stevenson's prose; he shares these preferences, but with a vanity that turns them into the attributes of an actor. It would be an exaggeration to say that our relationship is a hostile one; I live, I go on living, so that Borges may contrive his literature; and that literature justifies me.... Years ago I tried to free myself from him, and I went from mythologies of the city suburbs to

games with time and infinity, but now those games belong to Borges, and I will have to think up something else. Thus is my life a flight, and I lose everything, and everything belongs to oblivion, or to him. I don't know which of the two of us is writing this page.[2]

Borges is fleeing from himself, but of course he is also the pursuer. A similar image of self-pursuit turns up in a footnote in William James's *The Varieties of Religious Experience*. In the footnote, James quotes a philosopher named Xenos Clark describing a revelation induced in him by anesthesia. Clark emerged from the experience convinced that

> ordinary philosophy is like a hound hunting his own tail. The more he hunts, the farther he has to go, and his nose never catches up with his heels, because it is forever ahead of them. So the present is always a foregone conclusion, and I am ever too late to understand it. But at the moment of recovery from anesthesia, just then, before starting out on life, I catch, so to speak, a glimpse of my heels, a glimpse of the eternal process just in the act of starting. The truth is that we travel on a journey that was accomplished before we set out; and the real end of philosophy is accomplished, not when we arrive at, but when we remain in, our destination (being already there)—which may occur vicariously in this life when we cease our intellectual questioning.[3]

But we cannot cease our intellectual questioning. If we do, there is nothing. There is oblivion. The physicist John Wheeler, namer of black holes and prophet of the it from bit, intuited this truth. At the heart of reality lies not an answer, but a question: why is there something rather than nothing? *The Answer* is that there is no answer, only a question. Wheeler's suspicion that the world is nothing but "a figment of the imagination" was also well-founded. The world is a riddle that God has created in order to shield himself from his terrible solitude and fear of death.

Charles Hartshorne's Immortal God

I have sought, in vain, a theologian sympathetic to this terror-of-God idea. Freeman Dyson gave me one possible lead. After a lecture in which he had discussed his theological views—namely, the proposition that God is not omniscient or omnipotent but grows and learns as we humans grow and learn—Dyson was approached by an elderly man. The man identified himself as Charles Hartshorne, who, Dyson learned later, was among this century's most eminent theologians. Hartshorne told Dyson that his concept of God resembled that of a sixteenth-century Italian cleric named Socinus, who had been burned at the stake for heresy. Dyson got the impression that Hartshorne was a Socinian himself. I asked Dyson if he knew whether Hartshorne was still alive. "I'm not sure," Dyson replied. "He must be very, very old if he is still alive."

After leaving Princeton, I bought a book of essays on Hartshorne's theology, written by him and others, and found that he was indeed a Socinian.[4] Perhaps he would understand my terror-of-God idea—if he was alive. I looked in some back issues of *Who's Who* and found that Hartshorne's last position was at the University of Texas in Austin. I called the department of philosophy there and the secretary said yes, Professor Hartshorne was very much alive. He came in several times a week. But probably the best way to reach him was to call him at home.

Hartshorne answered the telephone. I identified myself and said I had received his name from Freeman Dyson. Did Hartshorne remember his conversation with Dyson on the topic of Socinus? Indeed he did. Hartshorne spoke about Dyson for a while and then launched into a discussion of Socinus. Although his voice was hoarse and quavery, he spoke with absolute self-assurance. Less than a minute into our conversation, his voice cracked and shifted into a Mickey Mouse falsetto, adding a further touch of surreality to our conversation.

Unlike most medieval and even modern theologians, Hartshorne told me, the Socinians believed that God changed, learned, and evolved through time, just as we humans do. "You see, the great classical tradition in medieval theology was to say that God is immutable," he explained. "The Socinians said, 'No that's wrong,' and they were absolutely right. To me that's just obvious."

So God is not omniscient? "God knows everything that there *is* to know," Hartshorne replied. "But there are no such things as future events. They can't be known until they happen." Any fool knows that, his tone implied.

If God could not see the future, was it possible that he could fear it? Could he fear his own death? "No!" Hartshorne shouted, and laughed at the absurdity of the notion. "We are born and we die," he said. "That's how we differ from God. It's not worth talking about God if God is born and dies. It's not worth talking about God if God has birth and death. God experiences *our* birth, but as *our* birth, not as God's birth, and he experiences *our* death."

I tried to explain that I was proposing not that God might actually die but only that he might *fear* death, that he might doubt his own immortality. "Oh," Hartshorne said, and I could practically see him shaking his head. "I have no interest in that."

I asked if Hartshorne had heard of the Omega Point theory. "Is that Teilhard de Chardin . . . ?" Yes, I replied, Teilhard de Chardin was the inspiration. The general idea is that superintelligent machines created by humans spread through the entire universe . . . "Yeah," Hartshorne interrupted contemptuously. "I'm not much interested in that. That's pretty fanciful."

I wanted to reply: *That's* fanciful, but all this Socinian nonsense isn't? Instead I asked if Hartshorne felt there would be any end to the evolution and learning of God. For the first time, he paused before answering. God, he said finally, is not a being but a "mode of becoming"; there was no beginning to this becoming, and there will be no end. Ever.

Wouldn't it be nice to think so.

The Fingernails of God

I have tried to describe my terror-of-God idea to various acquaintances, but I have not had much more success with them than I had with Hartshorne. One fellow science writer, a hyper-rational sort, listened patiently to my spiel without once smirking. "Let me see if I've got this straight," he said when I had mumbled into an impasse. "You're saying that everything really comes down to, like, God chewing his fingernails?" I thought about that for a moment, and then I nodded. Sure, why not. Everything comes down to God chewing his fingernails.

Actually, I think the terror-of-God hypothesis has much to recommend it. It suggests why we humans, even as we are compelled to seek truth, also shrink from it. Fear of truth, of *The Answer*, pervades our cultural scriptures, from the Bible through the latest mad-scientist movie. Scientists are generally thought to be immune to such uneasiness. Some are, or seem to be. Francis Crick, the Mephistopheles of materialism, comes to mind. So does the icy atheist Richard Dawkins, and Stephen Hawking, the cosmic joker. (Is there some peculiarity in British culture that produces scientists so immune to metaphysical anxiety?)

But for every Crick or Dawkins there are many more scientists who harbor a profound ambivalence concerning the notion of absolute truth. Like Roger Penrose, who could not decide whether his belief in a final theory was optimistic or pessimistic. Or Steven Weinberg, who equated comprehensibility with pointlessness. Or David Bohm, who was compelled both to clarify reality and to obscure it. Or Edward Wilson, who lusted after a final theory of human nature and was chilled by the thought that it might be attained. Or Marvin Minsky, who was so aghast at the thought of single-mindedness. Or Freeman Dyson, who insisted that anxiety and doubt are essential to existence. The ambivalence of these truth seekers toward final knowledge reflects the ambivalence of God—or the Omega Point, if you will—toward absolute knowledge of his own predicament.

Wittgenstein, in his prose poem *Tractatus Logico-Philosophicus*, intoned, "Not *how* the world is, is the mystical, but *that* it is."[5] True enlightenment, Wittgenstein knew, consists of nothing more than jaw-dropping dumbfoundment at the brute fact of existence. The ostensible goal of science, philosophy, religion, and all forms of knowledge is to transform the great "Hunh?" of mystical wonder into an even greater "Aha!" of understanding. But after one arrives at *The Answer*, what then? There is a kind of horror in thinking that our sense of wonder might be extinguished, once and for all time, by our knowledge. What, then, would be the purpose of existence? There would be none. The question mark of mystical wonder can never be completely straightened out, not even in the mind of God.

I have an inkling how this sounds. I like to think of myself as a rational person. I am fond, overly fond, of mocking scientists who take their own metaphysical fantasies too seriously. But, to paraphrase Marvin Minsky again, we all have many minds. My practical, rational mind tells me this terror-of-God stuff is delusional nonsense. But I have other minds. One glances at an astrology column now and then, or wonders if maybe there really *is* something to all those reports about people having sex with aliens. Another one of my minds believes that everything comes down to God chewing his fingernails. This belief even gives me a strange kind of comfort. Our plight is God's plight. And now that science—true, pure, empirical science—has ended, what else is there to believe in?

Loose Ends

As a science writer, I have always given great weight to mainstream opinion. The maverick defying the status quo makes for an entertaining story, but almost invariably he or she is wrong and the majority is right. The reception of *The End of Science* has thus placed me in an awkward position. It's not that I didn't expect, even hope, the book's message to be denounced. But I didn't foresee how wide-ranging and nearly unanimous the denunciations would be.

My end-of-science argument has been publicly repudiated by President Clinton's science advisor, the administrator of NASA, a dozen or so Nobel laureates and scores of less prominent critics in every continent except Antarctica. Even those reviewers who said they enjoyed the book usually took pains to distance themselves from its premise. "I do not buy [the book's] central thesis of limits and twilights," Natalie Angier testified toward the end of her exceedingly kind critique in the *New York Times Book Review*, June 30, 1996.

One might think that this sort of benign rebuke would nudge me toward self-doubt. But since my book's publication last June, I have become even more convinced that I am right and almost everyone else is wrong, which is generally a symptom of incipient madness. That is not to say that I have constructed an airtight case for my hypothesis. My book, like all books, was a compromise between ambition and the competing demands of family, publisher, employer, and so on. As I reluctantly surrendered the final draft to my editor, I was all too aware of ways I might have improved it. In this afterword, I hope to tie up some of the book's more obvious loose ends and to respond to points—reasonable and ridiculous—raised by critics.

Just Another "End of" Book

Perhaps the most common response to *The End of Science* was, "It's just another end-of-something-big book." Reviewers implied that my tract and others of its ilk—notably Francis Fukuyama's *End of History* and Bill McKibben's *End of Nature*—are manifestations of the same pre-millennial pessimism, a fad that need not be taken too seriously. Critics also accused my fellow end-ers and me of a kind of narcissism for insisting that ours is a special era, one of crises and culminations. As the *Seattle Times* put it, "We all want to live in a unique time; and proclamations of the end of history, a new age, a second coming, or the end of science are irresistible" (July 9, 1996).

But our age *is* unique. There is no precedent for the collapse of the Soviet Union, for a human population approaching six billion, for industry-induced global warming and ozone-depletion. There is certainly no precedent for thermonuclear bombs or moon landings or laptop computers or tests for breast-cancer genes—in short, for the explosion of knowledge and technology that has marked this century. Because we were all born into and grew up within this era, we simply assume that exponential progress is now a permanent feature of reality that will, must, continue. But a historical perspective suggests that such progress is probably an anomaly that will, must, end. Belief in the eternality of progress—not in crises and culminations—is the dominant delusion of our culture.

The June 17, 1996 issue of *Newsweek* proposed that my vision of the future represents a "failure of imagination." Actually, it is all too easy to imagine great discoveries just over the horizon. Our culture does it for us, with TV shows like *Star Trek* and movies like *Star Wars* and car advertisements and political rhetoric that promise us tomorrow will be very different from—and almost certainly better than—today. Scientists, and science journalists, too, are forever claiming that revolutions and breakthroughs and holy grails are imminent.

What I want people to imagine is this: What if there is no big thing over the horizon? What if what we have is basically what we are going to have? We are not going to invent superluminal, warp-drive spaceships that zip us to other galaxies or even other universes. We are not going to become infinitely wise or immortal through genetic engineering. We are not going to discover the mind of God, as the atheist Stephen Hawking put it.

What will be our fate, then? I suspect it will be neither mindless hedonism, as Gunther Stent prophesied in *The Coming of the Golden Age*, nor mindless battle, as Fukuyama warned in *The End of History*, but some combination of the two. We will continue to muddle along as we have been, oscillating between pleasure and misery, enlightenment and befuddlement, kindness and cruelty. It won't be heaven, but it won't be hell, either. In other words, the post-science world won't be all that different from our world.

Given my fondness for "Gotcha!" games, I guess it's only fair that a few critics have tried to force-feed me my own medicine. The *Economist* declared triumphantly in its July 20, 1996 issue that my end-of-science thesis is itself an example of ironic theorizing, since it is ultimately untestable and unprovable. But as Karl Popper replied when I asked him if his falsification hypothesis was falsifiable, "This is one of the most idiotic criticisms imaginable!" Compared to atoms or galaxies or genes or other objects of genuine scientific investigation, human culture is ephemeral; an asteroid could destroy us at any moment and bring about the end not only of science but also of history, politics, art—you name it. So *obviously* forecasts of human culture are educated guesses, at best, compared to predictions made in nuclear physics or astronomy or molecular biology, disciplines that address more permanent aspects of reality and can achieve more permanent truths. In that sense, yes, my end-of-science hypothesis is ironic.

But just because we cannot know our future with certainty does not mean we cannot make cogent arguments in favor of one future over another. And just as some works of philosophy or literary criticism or other ironic enterprises are more plausible than others, so are some predictions about the future of human culture. I think my scenario is more plausible than the ones I am trying to displace, in which we continue to discover profound new truths about the universe forever, or we arrive at an end point in which we achieve perfect wisdom and mastery over nature.

Is The End of Science *Antiscience?*

Over the past few years, scientists have bemoaned in ever-more-strident terms what they depict as a rising tide of irrationality and hostility toward science. The "antiscience" epithet has been hurled at targets as disparate as postmodern philosophers who challenge science's claim to absolute truth, Christian creationists, purveyors of occult schlock such as *The X Files*, and, not surprisingly, me. Philip Anderson, who won a Nobel prize for his work in condensed-matter physics, complained in the *London Times Higher Education Supplement* (September 27, 1996), that by treating certain scientists and theories so critically I have "most mischievously provided ammunition for the wave of antiscientism we are experiencing."

Ironically, particle physicists tarred Anderson with the antiscience brush because he criticized the Superconducting Supercollider before its cancellation in 1993. As Anderson responded when I asked him about his own reputation for judging other scientists harshly, "I just call 'em as I see 'em." I tried to portray the scientists and philosophers whom I interviewed for my book—and my reactions to them—as vividly and honestly as possible.

Let me reiterate what I said in my introduction: I became a science writer because I think science, and especially pure science, is the most miraculous and meaningful of all human creations. Moreover, I am not a Luddite. I am fond of my laptop computer, fax machine, television, and car. Although I deplore certain byproducts of science, such as pollution and nuclear weapons and racist theories of intelligence, I believe that science on the whole has made our lives immeasurably richer, intellectually and materially. On the other hand, science does not need another public-relations flak. In spite of its recent woes, science is still an immensely powerful force in our culture, much more so than postmodernism or creationism or other alleged threats. Science needs—and it can certainly withstand—informed criticism, which I humbly strive to provide.

Some observers have worried that *The End of Science* will be used to justify deep cuts in, if not the elimination of, funding for research. I might be concerned myself if there had been a groundswell of support for my thesis among federal officials, members of Congress, or the public. The opposite has happened. For the record, I do not advocate further decreases in funding for science, pure or applied, especially when the defense budget is still so obscenely large.

I take more seriously the complaint that my predictions might discourage young people from pursuing careers in science. The "inevitable corollary" of my argument, proclaimed the *Sacramento News* (July 18, 1996), is that "there's no point in trying to accomplish, see, experience anything new. We might as well kill all of our children." Well, I wouldn't go quite that far. Within this hysteria a legitimate point is buried, one that, because I have two tiny children of my own, I've given much thought to. What would I say to my own kids if they asked me whether I thought they should become scientists?

My answer might go something like this: Nothing I've written should discourage you from becoming a scientist. There are many tremendously important and exciting things left for scientists to do: finding better treatments for malaria or AIDS, less environmentally harmful sources of energy, more accurate forecasts of how pollution will affect climate. But if you want to discover something as monumental as natural selection or general relativity or the big bang theory, if you want to top Darwin or Einstein, your chances are slim to none. (Given their personalities, my kids might devote themselves to proving what an utter fool I am.)

In *Genius*, his biography of Richard Feynman, James Gleick pondered why science doesn't seem to produce giants like Einstein and Bohr anymore. The paradoxical answer, Gleick suggested, is that there are so *many* Einsteins and Bohrs—so many genius-level scientists—that it has become harder for any individual to stand apart from the pack. I'll buy that. But the crucial component missing from Gleick's hypothesis is that the geniuses of our era have less to discover than Einstein and Bohr did.

Returning for a moment to the antiscience issue, one of science's dirty little secrets is that many prominent scientists harbor remarkably postmodern sentiments. My book provides ample evidence of this phenomenon. Recall Stephen Jay Gould confessing his fondness for the seminal postmodern text *Structure of Scientific Revolutions*; Lynn Margulis declaring, "I don't think there's absolute truth, and if there is, I don't think any person has it"; Freeman Dyson predicting that modern physics will seem as primitive to future scientists as the phsyics of Aristotle seems to us.

What explains this skepticism? For these scientists as for many other intellectuals, truth-seeking, not truth itself, is what makes life meaningful, and to the extent that our current knowledge is true, it is that much more difficult to transcend. By insisting that our current knowledge may prove to be ephemeral, these skeptics can maintain the illusion that the great age of discovery is not over, that deeper revelations lie ahead. Postmodernism decrees that all future revelations will eventually prove to be ephemeral as well and will yield to other pseudo-insights, ad infinitum, but postmodernists are willing to accept this Sisyphean condition of their existence. They sacrifice the notion of absolute truth so that they can seek the truth forever.

A Point of Definition

On July 23, 1996, I appeared on the *Charlie Rose Show* with Jeremiah Ostriker, an astrophysicist from Princeton who was supposed to rebut my thesis. At one point, Ostriker and I squabbled over the dark-matter problem, which posits that stars and other luminous objects comprise only a small percentage of the total mass of the universe. Ostriker contended that the solution to the dark-matter problem would contradict my assertion that cosmologists would achieve no more truly profound discoveries; I disagreed, saying that the solution would turn out to be trivial. Our dispute, Rose interjected, seemed to involve just "a point of definition."

Rose touched on what I must admit is a shortcoming of my book. In arguing that scientists will not discover anything as fundamental as Darwin's theory of evolution or quantum mechanics, I should have spelled out more carefully what I meant by fundamental. A fact or theory is fundamental in proportion to how broadly it applies both in space and in time. Both quantum electrodynamics and general relativity apply, to the best of our knowledge, throughout the entire universe at all times since its birth. That makes these theories truly fundamental. A theory of high-temperature superconductivity, in contrast, applies only to specific types of matter that may exist, as far as we know, only in laboratories here on earth.

Inevitably, more subjective criteria also come into play in rankings of scientific findings. Technically, all biological theories are less fundamental than the cornerstone theories of physics, because biological theories apply—again, as far as we know—only to particular arrangements of matter that have existed on our lonely little planet for the past 3.5 billion years. But biology has the potential to be more *meaningful* than physics because it more directly addresses a phenomenon we find especially fascinating: ourselves.

In *Darwin's Dangerous Idea*, Daniel Dennett argued persuasively that evolution by natural selection is "the single best idea anyone has ever had," because it "unifies the realm of life, meaning, and purpose with the realm of space and time, cause and effect, mechanism and physical law." Indeed, Darwin's achievement—particularly when fused with Mendelian genetics into the new synthesis—has rendered all subsequent biology oddly anticlimactic, at least from a philosophical perspective (although, as I argue later, evolutionary biology offers limited insights into human nature). Even Watson and Crick's discovery of the double helix, although it has had enormous practical consequences, merely revealed the mechanical underpinnings of heredity; no significant revision of the new synthesis was required.

Returning to my dispute with Jeremiah Ostriker, my position is that cosmologists will never top the big bang theory, which holds basically that the universe is expanding and was once much smaller, hotter, and denser than it is today. The theory provides a coherent narrative, one with profound theological overtones, for the history of the universe. The universe had a beginning, and it might have an end (although cosmologists may never have enough evidence to settle this latter issue conclusively). What can be more profound, more meaningful, than that?

In contrast, the most likely solution to the dark-matter problem is relatively insignificant. The solution alleges that the motions of individual galaxies and of clusters of galaxies are best explained by assuming that the galaxies contain dust, dead stars, and other conventional forms of matter that cannot be detected through telescopes. There are more dramatic versions of the dark-matter problem, which postulate that as much as 99 percent of the universe consists of some exotic matter unlike anything we are familiar with here on earth. But these versions are predictions of inflation and other far-fetched cosmic suppositions that will never be confirmed, for reasons elaborated upon in Chapter Four.

What About Applied Science?

A couple of critics faulted me for neglecting—and implicitly denigrating—applied science. Actually, I think a good case can be made that applied science, too, is rapidly approaching its limits. For example, it once seemed inevitable that physicists' knowledge of nuclear fusion—which gave us the hydrogen bomb—would also yield a clean, economical, boundless source of energy. For decades, fusion researchers have said, "Keep the money coming, and in 20 years we will give you energy too cheap to meter." But in the last few years, the United States has drastically cut back on its fusion budget. Even the most optimistic researchers now predict that it will take at least 50 years to build economically viable fusion reactors. Realists acknowledge that fusion energy is a dream that may never be fulfilled: The technical, economic, and political obstacles are simply too great to overcome.

Turning to applied biology, its endpoint is nothing less than human immortality. The possibility that scientists can identify and then arrest the mechanisms underlying senescence is a perennial favorite of science writers. One might have more confidence in scientists' ability to crack the riddle of senescence if they had had more success with a presumably simpler problem: cancer. Since President Richard Nixon officially declared a federal "war on cancer" in 1971, the U.S. has spent some $30 billion on research, but cancer mortality rates have actually *risen* by 6 percent since then. Treatments have also changed very little. Physicians still cut cancer out with surgery, poison it with chemotherapy, and burn it with radiation. Maybe someday all our research will yield a "cure" that renders cancer as obsolete as smallpox. Maybe not. Maybe cancer—and by extension mortality—is simply too complex a problem to solve.

Ironically, biology's inability to defeat death may be its brightest hope. In the November/December 1995 issue of *Technology Review*, Harvey Sapolsky, a professor of social policy at MIT, noted that the major justification for the funding of science after World War II was national security—or, more specifically, the Cold War. Now that scientists no longer have the Evil Empire to justify their huge budgets, Sapolsky asked, what other opponent can serve as a substitute? The answer he came up with was mortality. Most people think living longer, and possibly even forever, is desirable, he pointed out. And the best thing about making immortality the primary goal of science, Sapolsky added, is that it is almost certainly unattainable, so scientists can keep getting funds for more research forever.

What About the Human Mind?

In a review in the July 1996 issue of *IEEE Spectrum*, the science writer David Lindley granted that physics and cosmology might well have reached dead ends. (This concession was not terribly surprising, given that Lindley wrote a book called *The End of Physics*.) But he nonetheless contended that investigations of the human mind—although now in a "prescientific state" in which scientists cannot even agree on what, precisely, they are studying—might eventually yield a powerful new paradigm. Maybe. But science's inability to move beyond the Freudian paradigm does not inspire much hope.

The science of mind has—in certain respects—become much more empirical and less speculative since Freud established psychoanalysis a century ago. We have acquired an amazing ability to probe the brain, with microelectrodes, magnetic resonance imaging, and positron-emission tomography. But this research has not led either to deep intellectual insights or to dramatic advances in treatment, as I tried to show in an article in the December 1996 issue of *Scientific American* called "Why Freud Isn't Dead." The reason psychologists, philosophers, and others still engage in protracted debates over Freud's work is that no undeniably superior theory of or therapy for the mind—either psychological or pharmacological—has emerged to displace psychoanalysis once and for all.

Some scientists think the best hope for a unifying paradigm of the mind may be Darwinian theory, which in its latest incarnation is called evolutionary psychology. Back in Chapter Six, I quoted Noam Chomsky's complaint that "Darwinian theory is so loose it can incorporate anything [scientists] discover." This point is crucial, and I elaborated on it in an article in the October 1995 issue of *Scientific American* called "The New Social Darwinists." The major counter-paradigm to evolutionary psychology is what could be called cultural determinism, which posits that culture rather than genetic endowment is the major molder of human behavior. To support their position, cultural determinists point to the enormous variety of behavior—much of it seemingly nonadaptive—displayed by people of different cultures.

In response, some evolutionary theorists have postulated that conformity—or "docility"—is an adaptive, innate trait. In other words, those who go along, get along. The Nobel laureate Herbert Simon conjectured in

Science (December 21, 1990) that docility could explain why people obey religious tenets that curb their sexuality or fight in wars when as individuals they often have little to gain and much to lose. While Simon's hypothesis cleverly coopts the culturalists' position, it also undermines the status of evolutionary psychology as a legitimate science. If a given behavior accords with Darwinian tenets, fine; if it does not, the behavior merely demonstrates our docility. The theory becomes immune to falsification—thus corroborating Chomsky's complaint that Darwinian theory can explain anything.

Acknowledging the tendency of humans to conform to their culture poses another problem for Darwinian theorists. To demonstrate that a given trait is innate, Darwinians try to show that it occurs in all cultures. In this way, for example, Darwinians have sought to show that males are inherently more inclined toward promiscuity than females. But given the interconnectedness of modern cultures, some of the universal and thus putatively innate attitudes and actions documented by Darwinian researchers might actually result from docility. That is what the cultural determinists have said all along.

Science's inability to grasp the mind is also reflected in the record of artificial intelligence, the effort to create computers that mimic human thought. Many pundits saw the chess match between the IBM computer "Deep Blue" and world champion Gary Kasparov in February 1996 as a triumph for artificial intelligence. After all, Deep Blue prevailed in the first game of the match before succumbing by a score of four points to two. To my jaundiced eye, the match underscored what a flop artificial intelligence has been since it was created by Marvin Minsky and others more than 40 years ago. Chess, with its straightforward rules and tiny Cartesian playing field, is a game tailor-made for computers. And Deep Blue, whose five human handlers include the best chess programmers in the world, is a prodigiously powerful machine, with 32 parallel processors capable of examining 200 million positions each second. If this silicon monster cannot defeat a mere human at chess, what hope is there that computers will ever mimic our more subtle talents, like recognizing a college sweetheart at a cocktail party and instantly thinking of just the right thing to say to make her regret dumping you 15 years ago?

The Chaoplexity Gambit

Since my book's original publication I have developed a couple of additional arguments about the limits of chaos and complexity—which I snidely lump together under the term *chaoplexity*. One of the most profound goals of chaoplexity—pursued by Stuart Kauffman, Per Bak, John Holland, and others—is the elucidation of a new law, or set of principles, or unified theory, or *something* that will make it possible to understand and predict the behavior of a wide variety of seemingly dissimilar complex systems. A closely related proposal is that the universe harbors a complexity-generating force that counteracts the second law of thermodynamics and creates galaxies, life, and even life intelligent enough to contemplate itself.

For such hypotheses to be meaningful, proponents must tell us what, exactly, complexity is and how it can be measured. We all sense intuitively that life today is more "complex" than it was 2,000, or 2,000,000, or 2,000,000,000 years ago, but how can that intuition be quantified in a nonarbitrary way? Until or unless this problem is solved, all these hypotheses about laws of complexity or complexity-generating forces are meaningless. I doubt (surprise, surprise) that the problem can be solved. Underlying most definitions of complexity is the notion that the complexity of a phenomenon is proportional to its improbability, or inversely proportional to its inevitability. If we shake a bag of molecules, how likely are we to get a galaxy, a planet, a paramecium, a frog, a stock broker? The best way to answer such questions would be to find other universes or other biological systems and analyze them statistically. That is obviously not possible.

Chaoplexologists nonetheless contend that they can answer these probability questions by constructing alternative universes and life histories in computers and determining which features are robust and which are contingent, or ephemeral. This hope stems, I believe, from an overly optimistic interpretation of certain developments in computer science and mathematics. Over the past few decades, researchers have found that various simple rules, when followed by a computer, can generate patterns that *appear* to vary randomly as a function of time or scale. Let's call this illusory randomness "pseudo-noise." The paradigmatic pseudo-noisy system is the Mandelbrot set, which has become an icon of the chaoplexity movement. The fields of both chaos and complexity have held out the hope that much of the noise that seems to pervade nature is actually pseudo-noise, the result of some underlying, deterministic algorithm.

But the noise that makes it so difficult to predict earthquakes, the stock market, the weather, and other phenomena is not apparent but very real, in my view, and will never be reduced to any simple set of rules. To be sure, faster computers and advanced mathematical techniques will improve our ability to predict certain complicated phenomena. Popular impressions notwithstanding, weather forecasting has become more accurate over the last few decades, in part because of improvements in computer modeling. But even more important are improvements in data-gathering—notably satellite imaging. Meteorologists have a larger, more accurate database upon which to build their models and against which to test them. Forecasts improve through this dialectic between simulation and data-gathering.

At some point, computer models drift over the line from science per se toward (shudder) engineering. The model either works or doesn't work according to some standard of effectiveness; "truth" is irrelevant. Moreover, chaos theory tells us that the butterfly effect imposes a fundamental limit on forecasting. One has to know the initial conditions of certain systems with infinite precision to be able to predict their course. This is something that has always puzzled me about chaoplexologists: According to one of their fundamental tenets, the butterfly effect, many of their goals may be impossible to achieve.

Chaoplexologists are not alone in addressing questions concerning the probability of various features of reality. These questions have also spawned such ironic hypotheses as the anthropic principle, inflation, multiuniverse theories, punctuated equilibrium, and Gaia. Unfortunately, you cannot determine the probability of the universe or of life on earth when you have only one universe and one history of life to contemplate. Statistics require more than one data point.

Lack of empirical data does not stop scientists and philosophers from holding strong opinions on these matters. On one side are inevitabilists, who take comfort in theories portraying reality as the highly probable and even necessary outcome of immutable laws. Most scientists are inevitabilists; perhaps the most prominent was Einstein, who rejected quantum mechanics because it implies that God plays dice with the universe. But there are some prominent anti-inevitabilists—notably Karl Popper, Stephen Jay Gould, and Ilya Prigogine—who see scientific determinism as a threat to human freedom and thus embrace uncertainty and randomness. We are either pawns of destiny or wildly improbable flukes. Take your pick.

Life on Mars?

In August 1996, two months after my book was published, scientists from NASA and elsewhere announced that they had discovered traces of fossilized microbial life in a meteorite fragment that had made its way from Mars to Antarctica. Commentators immediately seized upon this finding as proof of the absurdity of the suggestion that science might be ending. But far from disproving my thesis, the life-on-Mars story corroborates my assertion that science is in the throes of a great crisis. I am not cynical enough to believe what some observers have suggested, that NASA officials touted what they knew were flimsy findings in an effort to drum up more funds. But the hyperbolic response to the story—by NASA, politicians, the media, the public, and some scientists—demonstrates how desperate everyone is for a genuinely profound scientific discovery.

As I stated at several points in my book, the confirmation that we are not alone in the universe would represent one of the most thrilling events in human history. I hope to live long enough to witness such a revelation. But the finding of the NASA group doesn't come close. From the beginning, scientists who are truly knowledgeable about primordial biochemistry doubted that the life-on-Mars story would hold up. In December 1996, two groups of scientists independently reported in the journal *Geochimica et Cosmochimica Acta* that the alleged biological materials found in the Martian meteorite had probably been generated by nonbiological processes or by contamination from terrestrial organisms. "Death knell for Martian life," lamented the December 21/28, 1996 issue of *New Scientist.*

We will only know for certain whether life exists on Mars when or if we conduct a thorough search on the planet. Our best hope is for a human crew to drill deep below the surface, where there may be enough liquid water and heat to sustain microbial life as we know it. It will take decades, at least, for space officials to muster the money and technical resources for such a mission, even if politicians and the public are willing to pay for it.

Let's say that we do eventually determine that microbial life existed or still exists on Mars. That finding would provide an enormous boost for origin-of-life studies and biology in general. But would it mean that science is suddenly liberated from all its physical constraints? Hardly. If we find life on Mars, we will know that life exists elsewhere in this solar system. But we will be just as ignorant about whether life exists beyond our solar system, and we will still face huge obstacles to answering that question definitively.

Astronomers have recently identified a number of nearby stars orbited by planets, which may be capable of sustaining life. But Frank Drake, a physicist who was one of the founders of the Search for Extraterrestrial Intelligence program, called SETI, has estimated that current spacecraft would take *400,000 years* to reach the nearest of these planetary systems and establish whether they are inhabited. Someday, perhaps, the radio receivers employed in the SETI program will pick up electromagnetic signals—the alien equivalent of *I Love Lucy*—emanating from another star.

But as Ernst Mayr, one of this century's most eminent evolutionary biologists, has pointed out, most SETI proponents are physicists like Drake, who have an extremely deterministic view of reality. Physicists think that the existence of a highly technological civilization here on earth makes it highly probable that similar civilizations exist within signaling distance of earth. Biologists like Mayr find this view ludicrous, because they know how much contingency—just plain luck—is involved in evolution; rerun the great experiment of life a million times over and it might not produce mammals, let alone mammals smart enough to invent television. In an essay in the 1995 Cambridge University Press edition of *Extraterrestrials: Where Are They?*, Mayr concluded that the SETI program is bucking odds of "astronomical dimensions." Although I think Mayr is probably right, I was still dismayed when Congress terminated the funding for SETI in 1993. The program now limps along on private funds.

The Woo-Woo Stuff

Finally, there is the concluding section of my book, which veers into theology and mysticism, or what one acquaintance called "woo-woo stuff." I was worried that some reviewers would use this material to dismiss me—and thus my overall argument about the future of science—as irremediably flakey. Fortunately, that did not really happen. Most reviewers either ignored the epilogue or briefly expressed puzzlement over it.

The most astute—or should I say sympathetic?—interpretation was offered by the physicist Robert Park in the *Washington Post Book World*, August 11, 1996. He said that initially he was disappointed that I ended the book with "naive ironic science gone mad." But on further reflection he concluded that the ending was "a metaphor. This, Horgan is warning, is where science is ending. . . . Science has manned the battlements against the postmodern heresy that there is no objective truth, only to discover postmodernism inside the wall."

I couldn't have said it better myself. But I had other motives in mind as well. First, I felt it was only fair to reveal that I am as subject to metaphysical fantasies as those scientists whose views I mocked in the book. Also, the mystical episode I describe in the epilogue is the most important experience of my life. It had been burning a hole in my pocket, as it were, for more than 10 years, and I was determined to make use of it, even if it meant damaging what little credibility I may have as a journalist.

There is only one theological question that really matters: If there is a God, why has he created a world with so much suffering? My experience suggested an answer: If there is a God, he created the world out of terror and desperation as well as out of joy and love. This is my solution to the riddle of existence, and I had to share it. Let me be completely frank here. My real purpose in writing *The End of Science* was to found a new religion, "The Church of the Holy Horror." Being a cult leader should be a nice change of pace from—not to mention more lucrative than—science journalism.

New York, January 1997

Acknowledgments

I could never have written this book if *Scientific American* had not generously allowed and even encouraged me to pursue my own interests. *Scientific American* has also permitted me to adapt material from the following articles that I wrote for the magazine (copyright by *Scientific American*, Inc., all rights reserved): "Profile: Clifford Geertz," July 1989; "Profile: Roger Penrose," November 1989; "Profile: Noam Chomsky," May 1990; "In the Beginning," February 1991; "Profile: Thomas Kuhn," May 1991; "Profile: John Wheeler," June 1991; "Profile: Edward Witten," November 1991; "Profile: Francis Crick," February 1992; "Profile: Karl Popper," November 1992; "Profile: Paul Feyerabend," May 1993; "Profile: Freeman Dyson," August 1993; "Profile: Marvin Minsky," November 1993; "Profile: Edward Wilson," April 1994; "Can Science Explain Consciousness?" July 1994; "Profile: Fred Hoyle," March 1995; "From Complexity to Perplexity," June 1995; "Profile: Stephen Jay Gould," August 1995. I have also been granted permission to reprint excerpts from the following: *The Coming of the Golden Age*, by Gunther Stent, Natural History Press, Garden City, N.Y., 1969; *Scientific Progress*, by Nicholas Rescher, Blackwell, Oxford, U.K., 1978; *Farewell to Reason*, by Paul Feyerabend, Verso, London, 1987; and *Cosmic Discovery*, by Martin Harwit, MIT Press, Cambridge, 1984.

I am indebted to my agent, Stuart Krichevsky, for helping me to turn an amorphous idea into a marketable proposal, and to Bill Patrick and Jeff Robbins of Addison-Wesley, for providing just the right combination of criticism and encouragement. I am grateful to friends, acquaintances and colleagues at *Scientific American* and elsewhere who have given me feedback of various kinds, in some cases over a period of years. They include, in alphabetical order, Tim Beardsley, Roger Bingham, Chris Bremser, Fred Guterl, George Johnson, John Rennie, Phil Ross, Russell Ruthen, Gary Stix, Paul Wallich, Karen Wright, Robert Wright, and Glenn Zorpette. I owe thanks above all to my wife, Suzie, without whom I would have nothing.

NOTES

Introduction: Searching for The Answer

1. *The Emperor's New Mind*, Roger Penrose, Oxford University Press, New York, 1989. The review of the book, by the astronomer and author Timothy Ferris, appeared in the *New York Times Book Review* on November 19, 1989, p. 3.
2. My profile of Penrose was published in the November 1989 issue of *Scientific American*, pp. 30-33.
3. My meeting with Penrose in Syracuse took place in August 1989.
4. This definition of irony is based on that set forth by Northrop Frye in his classic work of literary theory, *Anatomy of Criticism*, Princeton University Press, Princeton, N.J., 1957.
5. *The Anxiety of Influence*, Harold Bloom, Oxford University Press, New York, 1973.
6. Ibid., p. 21.
7. Ibid., p. 22.

Chapter One: The End of Progress

1. The proceedings of the Gustavus Adolphus symposium were published as *The End of Science? Attack and Defense*, edited by Richard Q. Selve, University Press of America, Lanham, Md., 1992.
2. *The Coming of the Golden Age: A View of the End of Progress*, Gunther S. Stent, Natural History Press, Garden City, N.Y., 1969. See also Stent's contribution to Selve, *End of Science?*
3. See *The Education of Henry Adams*, Massachusetts Historical Society, Boston, 1918 (reprinted by Houghton Mifflin, Boston, 1961). Adams set forth his law of acceleration in chapter 34, written in 1904.

4. Stent, *Golden Age*, p. 94.

5. Ibid., p. 111.

6. Linus Pauling set forth his prodigious knowledge of chemistry in *The Nature of the Chemical Bond and the Structure of Molecules and Crystals*, published in 1939 and reissued in 1960 by Cornell University Press, Ithaca, N.Y. It remains one of the most influential scientific texts of all time. Pauling told me that he had solved the basic problems of chemistry almost a decade before his book was published. When I interviewed him in Stanford, California, in September 1992, Pauling said: "I felt that by the end of 1930, or even the middle, that organic chemistry was pretty well taken care of, and inorganic chemistry and minerology—except the sulfide minerals, where even now more work needs to be done." Pauling died on August 19, 1994.

7. Stent, *Golden Age*, p. 74.

8. Ibid., p. 115.

9. Ibid., p. 138.

10. I interviewed Stent in Berkeley in June 1992.

11. I found this dispiriting fact on page 371 of *Coming of Age in the Milky Way*, Timothy Ferris, Doubleday, New York, 1988. For a deflating retrospective of the U.S. manned space program, written for the 25th anniversary of the first lunar landing, see "25 Years Later, Moon Race in Eclipse," by John Nobel Wilford, *New York Times*, July 17, 1994, p. 1.

12. This pessimistic (optimistic?) view of senescence can be found in "Aging as the Fountain of Youth," chapter 8 of *Why We Get Sick: The New Science of Darwinian Medicine*, by Randolph M. Nesse and George C. Williams, Times Books, New York, 1994. Williams is one of the underacknowledged deans of modern evolutionary biology. See also his classic paper, "Pleiotropy, Natural Selection, and the Evolution of Senescence," *Evolution*, vol. 11, 1957, pp. 398–411.

13. Michelson's remarks have been passed down in several different versions. The one quoted here was published in *Physics Today*, April 1968, p. 9.

14. Michelson's decimal-point comment was erroneously attributed to Kelvin on page 3 of *Superstrings: A Theory of Everything?* edited by Paul C. Davies and Julian Brown, Cambridge University Press, Cambridge, U.K., 1988. This book is also notable for revealing that the Nobel

laureate Richard Feynman harbored a deep skepticism toward superstring theory.

15. Stephen Brush offered this analysis of physics at the end of the nineteenth century in "Romance in Six Figures," *Physics Today*, January 1969, p. 9.

16. See, for example, "The Completeness of Nineteenth-Century Science," by Lawrence Badash, *Isis*, vol. 63, 1972, pp. 48–58. Badash, a historian of science at the University of California at Santa Barbara, concluded (p. 58) that "the malaise of completeness was far from virulent . . . it was more a 'low-grade infection,' *but nevertheless very real*" [italics in original].

17. Daniel Koshland's essay, "The Crystal Ball and the Trumpet Call," and the special section on predictions that followed it can be found in *Science*, March 17, 1995. The legend of the nearsighted patent commissioner was reiterated by cybermagnate Bill Gates on page xiii of his 1995 bestseller *The Road Ahead*, cowritten with Nathan Myhrvold and Peter Rinearson, Viking, New York.

18. "Nothing Left to Invent," Eber Jeffery, *Journal of the Patent Office Society*, July 1940, pp. 479–481. I am indebted to the historian of science Morgan Sherwood of the University of California at Davis for locating Jeffery's article for me.

19. *The Idea of Progress*, J. B. Bury, Macmillan, New York, 1932. My summary of Bury's views is adapted from Stent, *Golden Age*.

20. *The Paradoxes of Progress*, Gunther S. Stent, W. H. Freeman, San Francisco, 1978, p. 27. This book contains several chapters from Stent's earlier book, *The Coming of the Golden Age*, plus new discussions of biology, morality, and the cognitive limits of science.

21. *Science: The Endless Frontier*, by Vannevar Bush, was reissued by the National Science Foundation, Washington, D.C., on its 40th birthday, in 1990.

22. I found this quote from Engels in *Scientific Progress*, by Nicholas Rescher, Basil Blackwell, Oxford, U.K., 1978, pp. 123–124. Rescher, a philosopher at the University of Pittsburgh, also provided several references to show that Engels's belief in the infinite potential of science had persisted among modern Marxists. See also the prologue to *Paradoxes of Progress*, in which Stent noted that the most critical review of *The Coming of the Golden Age* was that of a Soviet philoso-

pher, V. Kelle, who argued that science was eternal and that Stent's end-of-science thesis was a symptom of the decadence of capitalism.

23. Havel's remarks can be found in *Science and Anti-Science*, by Gerald Holton, Harvard University Press, Cambridge, 1993, pp. 175–176. Holton is a philosopher at Harvard University.

24. This view of Spengler's work is abstracted from Holton, *Science and Anti-Science*. In *Science and Anti-Science* as well as in other publications (including an essay in *Scientific American*, October 1995, p. 191), Holton has attempted to repudiate the notion that science is ending by appealing to the authority of Einstein, who often suggested that the search for scientific truths is eternal. It seems not to have occurred to Holton that Einstein's views reflected wishful thinking rather than a hard-nosed assessment of science's prospects. Holton has also suggested that those who think science is ending are in general opposed to science and rationality. Of course, modern predictions that science is approaching a culmination have come for the most part not from antirationalists, such as Havel, but from scientists, such as Steven Weinberg, Richard Dawkins, and Francis Crick, who believe that science is the supreme route to truth.

25. "Science: Endless Horizons or Golden Age?" Bentley Glass, *Science*, January 8, 1971, pp. 23–29. Glass, the retiring president of the American Association for the Advancement of Science (AAAS), had previously delivered this lecture at the annual meeting of the AAAS in Chicago on December 28, 1970.

26. "Milestones and Rates of Growth in the Development of Biology," Bentley Glass, *Quarterly Review of Biology*, March 1979, pp. 31–53.

27. My telephone interview with Glass took place in June 1994.

28. "Hard Times," Leo Kadanoff, *Physics Today*, October 1992, pp. 9–11.

29. My telephone interview with Kadanoff took place in August 1994.

30. Rescher, *Scientific Progress*, p. 37.

31. Ibid., p. 207. Although I disagree with Rescher's analysis of science's prospects, his books, *Scientific Progress* and *The Limits of Science*, University of California Press, Berkeley, 1984, are unparalleled sources of information for anyone interested in the limits of science. Both books, unfortunately, are out of print.

32. Bentley Glass's review of Rescher's *Scientific Progress* was published in the *Quarterly Review of Biology*, December 1979, pp. 417–419.

33. I have lifted this remark by Kant from Rescher's *Scientific Progress*, p. 246.

34. The meaning of Bacon's phrase *plus ultra* is discussed in *The Limits of Science*, by Peter Medawar, Oxford University Press, New York, 1984. Medawar was a prominent British biologist.

35. *Critical Theory Since Plato*, edited by Hazard Adams, Harcourt Brace Jovanovich, New York, 1971, p. 474.

Chapter Two: The End of Philosophy

1. "Where Science Has Gone Wrong," T. Theocharis and M. Psimopoulos, *Nature*, vol. 329, October 15, 1987, pp. 595–598.

2. Peirce's view of the relationship between science and final truth is discussed in Rescher's *The Limits of Science* (see note 31 to Chapter 1). See also Peirce's *Selected Writings*, edited by Philip Wiener, Dover Publications, New York, 1966.

3. Popper's major works include *The Logic of Scientific Discovery*, Springer, Berlin, 1934 (reprinted by Basic Books, New York, 1959); *The Open Society and Its Enemies*, Routledge, London, 1945 (reprinted by Princeton University Press, Princeton, N.J., 1966); and *Conjectures and Refutations*, Routledge, London, 1963 (reprinted by Harper and Row, New York, 1968). Popper's autobiography, *Unended Quest*, Open Court, La Salle, Ill., 1985, and *Popper Selections*, edited by David Miller, Princeton University Press, Princeton, N.J., 1985, provide excellent introductions to his thought.

4. See "Who Killed Logical Positivism?" chapter 17 of Popper's *Unended Quest*.

5. Ibid., p. 116.

6. My interview with Popper took place in August 1992.

7. My article on quantum mechanics, "Quantum Philosophy," was published in *Scientific American*, July 1992, pp. 94–103.

8. See *The Self and Its Brain*, by Popper and John C. Eccles, Springer-Verlag, Berlin, 1977. Eccles won a Nobel Prize in 1963 for his work in neural signaling. I discuss his views in Chapter 7.

9. Günther Wächtershäuser has set forth his theory of the origin of life in the *Proceedings of the National Academy of Sciences*, vol. 87, 1990, pp. 200–204.

10. Popper discussed his doubts about Darwin's theory in "Natural Selection and Its Scientific Status," chapter 10 of *Popper Selections*.

11. Mrs. Mew was looking for Popper's book *A World of Propensities*, Routledge, London, 1990.

12. *Nature* published an homage to Popper by the physicist Hermann Bondi on July 30, 1992, p. 363. The occasion was the philosopher's 90th birthday.

13. The *Economist* published its obituary of Popper on September 24, 1994, p. 92. Popper had died on September 17.

14. Popper, *Unended Quest*, p. 105.

15. *The Structure of Scientific Revolutions*, Thomas Kuhn, University of Chicago Press, Chicago, 1962. (Page numbers refer to the 1970 edition.) My interview with Kuhn took place in February 1991.

16. *Scientific American*, May 1964, pp. 142–144.

17. Kuhn's comparison of scientists to addicts and to the brainwashed characters in *1984* can be found on pages 38 and 167 of *Structure*.

18. I originally made this snide remark about the Bush administration's New Paradigm in a profile of Kuhn in *Scientific American*, May 1991, pp. 40–49. Later I received a letter of complaint from James Pinkerton, who was then deputy assistant to President Bush for policy planning and had coined the term *New Paradigm*. Pinkerton insisted that the New Paradigm was "*not* a rehashing of Reaganomics; instead it is a coherent set of ideas and principles that emphasize choice, empowerment, and accomplishing more with less centralized control."

19. The charge that Kuhn defined paradigm in 21 different ways can be found in "The Nature of a Paradigm," by Margaret Masterman, in *Criticism and the Growth of Knowledge*, edited by Imre Lakatos and Alan Musgrave, Cambridge University Press, New York, 1970.

20. *Against Method*, Paul Feyerabend, Verso, London, 1975 (reprinted in 1993).

21. The "positivistic teacup" remark can be found in *Farewell to Reason*, by Paul Feyerabend, Verso, London, 1987, p. 282.

22. Feyerabend's organized crime analogy can be found in his essay "Consolations for a Specialist," in Lakatos and Musgrave, *Growth of Knowledge*.

23. Feyerabend's outrageous utterances were recounted in a surprisingly sympathetic profile by William J. Broad, now a science reporter for the *New York Times*: "Paul Feyerabend: Science and the Anarchist," *Science*, November 2, 1979, pp. 534–537.

24. Feyerabend, *Farewell to Reason*, p. 309.

25. Ibid., p. 313.
26. *Isis*, vol. 2, 1992, p. 368.
27. Feyerabend died on February 11, 1994, in Geneva. The *New York Times* ran his obituary on March 8.
28. *Killing Time*, Paul Feyerabend, University of Chicago Press, Chicago, 1995.
29. *After Philosophy: End or Transformation?* edited by Kenneth Baynes, James Bohman, and Thomas McCarthy, MIT Press, Cambridge, 1987.
30. *Problems in Philosophy*, Colin McGinn, Blackwell Publishers, Cambridge, Mass., 1993.
31. "The Zahir" can be found in *A Personal Anthology*, by Jorge Luis Borges, Grove Press, New York, 1967. This collection also contains two other chilling stories about absolute knowledge: "Funes, the Memorious" and "The Aleph."
32. Ibid., p. 137.

Chapter Three: The End of Physics

1. Einstein's remark can be found in *Theories of Everything*, by John Barrow, Clarendon Press, Oxford, U.K., 1991, p. 88.
2. Glashow's full remarks are reprinted in *The End of Science? Attack and Defense*, edited by Richard Q. Selve, University Press of America, Lanham, Md., 1992.
3. "Desperately Seeking Superstrings," Sheldon Glashow and Paul Ginsparg, *Physics Today*, May 1986, p. 7.
4. "A Theory of Everything," K. C. Cole, *New York Times Magazine*, October 18, 1987, p. 20. This article provided me with most of the personal information on Witten in this chapter. I interviewed Witten in August 1991.
5. See *Science Watch* (published by the Institute for Scientific Information, Philadelphia, Pa.), September 1991, p. 4.
6. Barrow, *Theories of Everything*.
7. *The End of Physics*, David Lindley, Basic Books, New York, 1993.
8. See "Is the *Principia* Publishable Now?" by John Maddox, *Nature*, August 3, 1995, p. 385.
9. *Lonely Hearts of the Cosmos*, Dennis Overbye, HarperCollins, New York, 1992, p. 372.
10. *Dreams of a Final Theory*, Steven Weinberg, Pantheon, New York, 1992, p. 18.

11. *The First Three Minutes*, Steven Weinberg, Basic Books, New York, 1977, p. 154.

12. Weinberg, *Dreams of a Final Theory*, p. 253.

13. *Hyperspace*, Michio Kaku, Oxford University Press, New York, 1994.

14. *The Mind of God*, Paul C. Davies, Simon and Schuster, New York, 1992. The judges who awarded Davies the Templeton Prize included George Bush and Margaret Thatcher.

15. Bethe first publicly discussed his fateful calculation in "Ultimate Catastrophe?" *Bulletin of the Atomic Scientists*, June 1976, pp. 36–37. The article is reprinted in a collection of Bethe's papers, *The Road from Los Alamos*, American Institute of Physics, New York, 1991. I interviewed Bethe at Cornell in October 1991.

16. "What's Wrong with Those Epochs?" David Mermin, *Physics Today*, November 1990, pp. 9–11.

17. Wheeler's essays and papers have been collected in *At Home in the Universe*, American Institute of Physics Press, Woodbury, N.Y., 1994. I interviewed Wheeler in April 1991.

18. See page 5 of Wheeler's essay "Information, Physics, Quantum: The Search for Links," in *Complexity, Entropy, and the Physics of Information*, edited by Wojciech H. Zurek, Addison-Wesley, Reading, Mass., 1990.

19. Ibid., p. 18.

20. This quotation, and the preceding story about Wheeler's appearing with parapsychologists at the American Association for the Advancement of Science meeting, can be found in "Physicist John Wheeler: Retarded Learner," by Jeremy Bernstein, *Princeton Alumni Weekly*, October 9, 1985, pp. 28–41.

21. For a concise introduction to Bohm's career, see "Bohm's Alternative to Quantum Mechanics," by David Albert, *Scientific American*, May 1994, pp. 58–67. Portions of this section on Bohm appeared in my article "Last Words of a Quantum Heretic," *New Scientist*, February 27, 1993, pp. 38–42. Bohm set forth his philosophy in *Wholeness and the Implicate Order*, Routledge, New York, 1983 (first printed in 1980).

22. The Einstein–Podolsky–Rosen paper, Bohm's original paper on his alternative interpretation of quantum mechanics, and many other seminal articles on quantum mechanics can be found in *Quantum Theory and Measurement*, edited by John Wheeler and Wojciech H. Zurek, Princeton University Press, Princeton, N.J., 1983.

23. *Science, Order, and Creativity*, David Bohm and F. David Peat, Bantam Books, New York, 1987.

24. I interviewed Bohm in August 1992. He died on October 27. Before his death he cowrote another book setting forth his views, which was published two years later, *The Undivided Universe*, by Bohm and Basil J. Hiley, Routledge, London, 1994.

25. *The Character of Physical Law*, Richard Feynman, MIT Press, Cambridge, 1967, p. 172. (Feynman's book was first published in 1965 by the BBC.)

26. Ibid., p. 173.

27. The symposium, "The Interpretation of Quantum Theory: Where Do We Stand?" took place at Columbia University, April 1–4, 1992.

28. I have seen many versions of this quote from Bohr. Mine comes from an interview with John Wheeler, who studied under Bohr.

29. For an excellent analysis of the state of physics, see "Physics, Community, and the Crisis in Physical Theory," by Silvan S. Schweber, *Physics Today*, November 1993, pp. 34–40. Schweber, a distinguished historian of physics at Brandeis University, suggests that physics will increasingly be directed toward utilitarian goals rather than toward knowledge for its own sake. I wrote about the difficulties that face physicists who are trying to achieve a unified theory in "Particle Metaphysics," *Scientific American*, February 1994, pp. 96–105. In an earlier article for *Scientific American*, "Quantum Philosophy," July 1992, pp. 94–103, I reviewed current work on the interpretation of quantum mechanics.

Chapter Four: The End of Cosmology

1. Hawking's lecture and the other proceedings of the Nobel symposium, which was held June 11–16, 1990, in Gräftvallen, Sweden, were published as *The Birth and Early Evolution of Our Universe*, edited by J. S. Nilsson, B. Gustafsson, and B.-S. Skagerstam, World Scientific, London, 1991. I also wrote an article based on the meeting, "Universal Truths," *Scientific American*, October 1990, pp. 108–117. I had a disturbing encounter with Stephen Hawking on my first day at the Nobel symposium, when all the participants of the meeting were herded into the woods for a cocktail party. We were within sight of the tables bearing food and drink when Hawking's wheelchair, which was being pushed by one of his nurses, jammed in a rut in the path. The nurse asked if I would mind carrying Hawking the rest of the way

to the party. Hawking, when I scooped him up, was disconcertingly light and stiff, like a bundle of sticks. I glanced at him out of the corner of my eye and found him already eyeing me suspiciously. Abruptly, his face twisted into an agonized grimace; his body shuddered violently, and he emitted a gargling noise. My first thought was: A man is dying in my arms! How horrible! My second thought was: Stephen Hawking is dying in my arms! What a story! That thought was surrendering in turn to shame at the depths of my opportunism when the nurse, who had noticed Hawking's distress, and mine, hustled up to us. "Don't worry," she said, gathering Hawking gently into her arms. "This happens to him all the time. He'll be all right."

2. A condensed version of Hawking's lecture, which he delivered on April 29, 1980, was published in the British journal *Physics Bulletin* (now called *Physics World*), January 1981, pp. 15–17.

3. *A Brief History of Time*, Stephen Hawking, Bantam Books, New York, 1988, p. 175.

4. Ibid., p. 141.

5. *Stephen Hawking: A Life in Science*, Michael White and John Gribbon, Dutton, New York, 1992. This book also documents Hawking's transformation from a physicist into an international celebrity.

6. See the interview with Hawking in *Science Watch*, September 1994. Hawking's views on the end of physics are discussed in several books cited in Chapter 3, including *The Mind of God*, by Paul C. Davies; *Theories of Everything*, by John Barrow; *Dreams of a Final Theory*, by Steven Weinberg; *Lonely Hearts of the Cosmos*, by Dennis Overbye; and *The End of Physics*, by David Lindley. See also *Fire in the Mind*, by George Johnson, Alfred A. Knopf, New York, 1995, for a particularly subtle discussion of whether science can attain absolute truth.

7. Schramm's mainstream view of cosmology is set forth in *The Shadows of Creation*, by Schramm and Michael Riordan, W. H. Freeman, New York, 1991. In 1994, Schramm's coauthor, Riordan, a physicist at the Stanford Linear Accelerator, bet me a case of California wine that by the end of the century Alan Guth, who is generally credited with having "discovered" inflation, would win a Nobel Prize for his work. I mention this bet here only because I am sure that I will win it.

8. Linde set forth his theory in "The Self-Reproducing Inflationary Universe," *Scientific American*, November 1994, pp. 48–55. Those who want more on Linde can sample his books *Particle Physics and Inflationary Cosmology*, Harwood Academic Publishers, New York, 1990;

and *Inflation and Quantum Cosmology*, Academic Press, San Diego, 1990. Portions of this section on Linde appeared in my article "The Universal Wizard," *Discover*, March 1992, pp. 80–85. My interview with Linde at Stanford took place in April 1991.

9. I spoke to Schramm by telephone in February 1993.

10. I interviewed Georgi at Harvard in November 1993.

11. Hoyle provided a charming retrospective of his tumultuous career in *Home Is Where the Wind Blows*, University Science Books, Mill Valley, Calif., 1994. I interviewed Hoyle at his home in August 1992.

12. See, for example, the review in *Nature*, May 13, 1993, p. 124, of *Our Place in the Cosmos*, J. M. Dent, London, 1993, in which Hoyle and his collaborator, Chandra Wickramasinghe, argue that the cosmos is teeming with life. *Nature*'s reviewer, Robert Shapiro, a chemist at New York University, asserted that this book and other recent ones by Hoyle "afford full documentation of the way in which a brilliant mind can be turned to the pursuit of bizarre ideas." When Hoyle's autobiography was published a year later, the media, which had for years marginalized Hoyle for his maverick views, showed a sudden fondness for him. See, for example, "The Space Molecule Man," Marcus Chown, *New Scientist*, September 10, 1994, pp. 24–27.

13. "And the Winner Is . . . ," *Sky and Telescope*, March 1994, p. 22.

14. See Hoyle and Wickramasinghe, *Our Place in the Cosmos*.

15. See Overbye, *Lonely Hearts of the Cosmos*, for an excellent account of the debate over the Hubble constant.

16. "The Scientist as Rebel," Freeman Dyson, *New York Review of Books*, May 25, 1995, p. 32.

17. *Cosmic Discovery*, Martin Harwit, MIT Press, Cambridge, 1981, pp. 42–43. In 1995, Harwit resigned from his job as director of the Smithsonian Institution's National Air and Space Museum in Washington, D.C., in the midst of a bitter controversy over an exhibit he had supervised, called "The Last Act: The Atomic Bomb and the End of World War II." Veterans and others had complained that the exhibit was too critical of the U.S. decision to drop atomic bombs on Hiroshima and Nagasaki.

18. Ibid., p. 44.

19. I found this quote from Donne at the end of an essay by the biologist Loren Eisley, "The Cosmic Prison," *Horizon*, Autumn 1970, pp. 96–101.

Chapter Five: The End of Evolutionary Biology

1. See the 1964 Harvard University Press edition of *On the Origin of Species*, with a foreword by Ernst Mayr, one of the founders of modern evolutionary theory.

2. Stent, *Golden Age*, p. 19.

3. I found this remark by Bohr in a book review in *Nature*, August 6, 1992, p. 464. The exact quote is: "It is the task of science to reduce deep truths to trivialities."

4. The gathering with Dawkins took place in November 1994 at the office of John Brockman, a spectacularly successful agent and public relations expert for scientist-authors.

5. *The Blind Watchmaker*, Richard Dawkins, W. W. Norton, New York, 1986, p. ix. The Wallace referred to by Dawkins is Alfred Russell Wallace, who discovered the concept of natural selection independently of Darwin but never approached Darwin's depth or breadth of insight.

6. See "Is Uniformitarianism Necessary?" *American Journal of Science*, vol. 263, 1965, pp. 223–228.

7. "Punctuated Equilibria: An Alternative to Phyletic Gradualism," Stephen Jay Gould and Niles Eldredge, in *Models in Paleobiology*, edited by T. J. M. Schopf, W. H. Freeman, San Francisco, 1972.

8. My favorites among Gould's many books are *The Mismeasure of Man*, W. W. Norton, New York, 1981, both a scholarly history of and an impassioned polemic against intelligence tests, and *Wonderful Life*, W. W. Norton, New York, 1989, a masterful exposition of his view of life as the product of contingency. See also "The Spandrels of San Marco and the Panglossian Paradigm," by Gould and his Harvard colleague Richard Lewontin (a geneticist who, like Gould, is often accused of Marxist leanings), in the *Proceedings of the Royal Society* (London), vol. 205, 1979, pp. 581–598. The article was a devastating critique of simplistic Darwinian explanations of physiology and behavior. For an equally sharp attack on Gould's evolutionary outlook, see the review of *Wonderful Life* by Robert Wright in the *New Republic*, January 29, 1990.

9. "Punctuated Equilibrium Comes of Age," *Nature*, November 18, 1993, pp. 223–227. I interviewed Gould in New York City in November 1994.

10. As reprinted in Dawkins, *Blind Watchmaker*, p. 245. The chapter

within which this quotation is embedded, "Puncturing Punctuationism," delivers on its title.

11. For a no-nonsense treatment of Lynn Margulis's work on symbiosis, see her book *Symbiosis in Cell Evolution*, W. H. Freeman, New York, 1981.

12. See Margulis's contributions to *Gaia: The Thesis, the Mechanisms, and the Implications*, edited by P. Bunyard and E. Goldsmith, Wadebridge Ecological Center, Cornwall, U.K., 1988.

13. *What Is Life?*, Lynn Margulis and Dorion Sagan, Peter Nevraumont, New York, 1995 (distributed by Simon and Schuster). I interviewed Margulis in May 1994.

14. This claim about Lovelock's crisis of faith can be found in "Gaia, Gaia: Don't Go Away," by Fred Pearce, *New Scientist*, May 28, 1994, p. 43.

15. This and other patronizing comments about Margulis can be found in "Lynn Margulis: Science's Unruly Earth Mother," by Charles Mann, *Science*, April 19, 1991, p. 378.

16. The works by Kauffman referred to in this section include "Antichaos and Adaptation," *Scientific American*, August 1991, pp. 78–84; *The Origins of Order*, Oxford University Press, New York, 1993; and *At Home in the Universe*, Oxford University Press, New York, 1995.

17. See my article "In the Beginning," *Scientific American*, February 1991, p. 123.

18. Brian Goodwin set forth his theory in *How the Leopard Changed Its Spots*, Charles Scribner's Sons, New York, 1994.

19. John Maynard Smith's disparaging remarks about the work of Per Bak and Stuart Kauffman were reported in *Nature*, February 16, 1995, p. 555. See also the insightful review of Kauffman's *Origins of Order* in *Nature*, October 21, 1993, pp. 704–706.

20. My February 1991 article in *Scientific American* (see note 17) reviewed the most prominent theories of the origin of life. I interviewed Stanley Miller at the University of California at San Diego in November 1990 and again by telephone in September 1995.

21. Stent, *Golden Age*, p. 71.

22. Crick's "miracle" comment can be found on page 88 of his book *Life Itself*, Simon and Schuster, New York, 1981.

Chapter Six: The End of Social Science

1. I interviewed Edward Wilson at Harvard in February 1994. The works by Wilson alluded to this section include *Sociobiology*, Harvard

University Press, Cambridge, 1975 (my citations refer to the abridged 1980 edition); *On Human Nature*, Harvard University Press, Cambridge, 1978; *Genes, Mind, and Culture* (with Charles Lumsden), Harvard University Press, Cambridge, 1981; *Promethean Fire* (with Lumsden), Harvard University Press, Cambridge, 1983; *Biophilia*, Harvard University Press, Cambridge, 1984; *The Diversity of Life*, W. W. Norton, New York, 1993; and *Naturalist*, Island Press, Washington, D.C., 1994.

2. See "The Molecular Wars," chapter 12 of *Naturalist*, for a detailed account of this crisis in Wilson's career.

3. Wilson, *Sociobiology*, p. 300.

4. These travails are recounted in the chapter titled "The Sociobiology Controversy," in Wilson's *Promethean Fire*, pp. 23–50.

5. Wilson, *Promethean Fire*, pp. 48–49.

6. Christopher Wells, a biologist at the University of California at San Diego, made these remarks about the theories of Wilson and Lumsden in *The Sciences*, November/December 1993, p. 39.

7. My own view is that scientists have not been nearly as successful at explaining human behavior in genetic and Darwinian terms as Wilson seemed to believe. See my *Scientific American* articles "Eugenics Revisited," June 1993, pp. 122–131; and "The New Social Darwinists," October 1995, pp. 174–181.

8. Wilson, *Sociobiology*, pp. 300–301.

9. See *One Long Argument*, by Ernst Mayr, Harvard University Press, Cambridge, 1991. On page 149 Mayr wrote: "The architects of the evolutionary synthesis [of whom Mayr was one] have been accused by some critics of claiming that they had solved all the remaining problems of evolution. This accusation is quite absurd; I do not know of a single evolutionist who would make such a claim. All that was claimed by the supporters of the synthesis was that they had arrived at an elaboration of the Darwinian paradigm sufficiently robust not to be endangered by remaining puzzles." "Puzzles," it should be recalled, is the term that Thomas Kuhn employed to describe problems that occupy scientists involved in nonrevolutionary, "normal" science.

10. I found this quotation from Darwin's *The Descent of Man* in *The Moral Animal*, by Robert Wright, Pantheon, New York, 1994, p. 327. (Wright cites the facsimile edition of *Descent*, Princeton University Press, Princeton, N.J., 1981, p. 73.) This book by Wright, a journalist associ-

ated with the *New Republic*, is by far the best I have read on recent scientific attempts to explain human nature in Darwinian terms.

11. Ibid., p. 328.

12. I met with Chomsky at MIT in February 1990. The remarks quoted up to this point came from that meeting. The remarks quoted hereafter stemmed from a telephone interview in February 1993. Chomsky's political essays have been collected in *The Chomsky Reader*, edited by James Peck, Pantheon, New York, 1987.

13. *The New Encyclopaedia Britannica*, 1992 *Macropaedia* edition, vol. 23, *Linguistics*, p. 45.

14. *Nature*, February 19, 1994, p. 521.

15. *Syntactic Structures*, Noam Chomsky, Mouton, The Hague, Netherlands, 1957. In 1995, Chomsky issued yet another book on linguistics, *The Minimalist Program*, MIT Press, Cambridge, which extended his earlier work on an innate, generative grammar. Like most of Chomsky's books on linguistics, this one is not easy to read. For a readable account of Chomsky's career in linguistics, see *The Linguistics Wars*, by Randy Allen Harris, Oxford University Press, New York, 1993.

16. Steven Pinker, who is also a linguist at MIT, nonetheless argued persuasively that Chomsky's work is best understood from a Darwinian viewpoint in *The Language Instinct*, William Morrow, New York, 1994.

17. *Language and the Problems of Knowledge*, Noam Chomsky, MIT Press, Cambridge, 1988, p. 159. Chomsky also spelled out his views on cognitive limits in this book.

18. Stent, *Golden Age*, p. 121.

19. "Thick Description: Toward an Interpretive Theory of Culture" can be found in Geertz's collection of essays, *The Interpretation of Cultures*, Basic Books, New York, 1973.

20. Ibid., p. 29.

21. *Works and Lives: The Anthropologist as Author*, Clifford Geertz, Stanford University Press, Stanford, 1988, p. 141.

22. "Deep Play" was collected in *The Interpretation of Cultures*. This quotation is found on page 412.

23. Ibid., p. 443.

24. I interviewed Geertz in person at the Institute for Advanced Study in May 1989 and again by telephone in August 1994.

25. *After the Fact*, Clifford Geertz, Harvard University Press, Cambridge, 1995, pp. 167–168.

Chapter Seven: The End of Neuroscience

1. Crick has written an illuminating account of his career: *What Mad Pursuit*, Basic Books, New York, 1988. He spelled out his views on consciousness in *The Astonishing Hypothesis*, Charles Scribner's Sons, New York, 1994.

2. "Toward a Neurobiological Theory of Consciousness," Francis Crick and Christof Koch, *Seminars in the Neurosciences*, vol. 2, 1990, pp. 263–275.

3. I interviewed Crick at the Salk Institute in November 1991.

4. *The Double Helix*, James Watson, Atheneum, New York, 1968.

5. Crick, *What Mad Pursuit*, p. 9.

6. Crick, *Astonishing Hypothesis*, p. 3.

7. Edelman's books on the mind, published by Basic Books, New York, include *Neural Darwinism*, 1987; *Topobiology*, 1988; *The Remembered Present*, 1989; and *Bright Air, Brilliant Fire*, 1992. All these books, even the final one, which was intended to set forth Edelman's views in a popular format, are extremely difficult to read.

8. "Dr. Edelman's Brain," Steven Levy, *New Yorker*, May 2, 1994, p. 62.

9. "Plotting a Theory of the Brain," David Hellerstein, *New York Times Magazine*, May 22, 1988, p. 16.

10. See Crick's stinging review of Edelman's book *Neural Darwinism*: "Neural Edelmanism," *Trends in Neurosciences*, vol. 12, no. 7, 1989, pp. 240–248.

11. Daniel Dennett reviewed *Bright Air, Brilliant Fire* in *New Scientist*, June 13, 1992, p. 48.

12. Koch made this remark at "Toward a Scientific Basis for Consciousness," a meeting held in Tucson, Arizona, April 12–17, 1994.

13. Eccles has set forth his views in various publications, including *The Self and Its Brain*, cowritten with Karl Popper, Springer-Verlag, Berlin, 1977; *How the Self Controls Its Brain*, Springer-Verlag, Berlin, 1994; "Quantum Aspects of Brain Activity and the Role of Consciousness," cowritten with Friedrich Beck, *Proceedings of the National Academy of Science*, vol. 89, December 1992, pp. 11,357–11,361. My interview with Eccles took place by telephone in February 1993.

14. I interviewed Roger Penrose at the University of Oxford in August 1992. Penrose's two books on consciousness are *The Emperor's New Mind*, Oxford University Press, New York, 1989; and *Shadows of the Mind*, Oxford University Press, New York, 1994.

15. For criticism of *The Emperor's New Mind*, see *Behavioral and Brain Sciences*, vol. 13, no. 4, December 1990. This issue has multiple reviews of Penrose's book. For sharp critiques of *Shadows of the Mind*, see "Shadows of Doubt," by Philip Anderson (a prominent physicist), *Nature*, November 17, 1994, pp. 288–289; and "The Best of All Possible Brains," by Hilary Putnam (a prominent philosopher), *New York Times Book Review*, November 20, 1994, p. 7.

16. *The Science of the Mind*, Owen Flanagan, MIT Press, Cambridge, 1991. Daniel Dennett drew my attention to Flanagan's term.

17. "What Is It Like to Be a Bat," by Thomas Nagel, can be found in *Mortal Questions*, Cambridge University Press, New York, 1979, a collection of Nagel's essays. This quotation is from page 166. In June 1992, I called Nagel to ask whether he thought science could ever end. Absolutely not, he replied. "The more you discover, the more questions there will be," he said. "Shakespearean criticism can never be complete," he added, "so why should physics be?"

18. I interviewed McGinn in New York City in August 1994. See McGinn's book *The Problem of Consciousness*, Blackwell Publishers, Cambridge, Mass., 1991, for a full discussion of his mysterian viewpoint.

19. *Consciousness Explained*, Daniel Dennett, Little, Brown, Boston, 1991. See also "The Brain and Its Boundaries," *London Times Literary Supplement*, May 10, 1991, in which Dennett attacks McGinn's mysterian position. I spoke to Dennett about the mysterian viewpoint by telephone in April 1994.

20. "Toward a Scientific Basis for Consciousness" took place in Tucson, Arizona, April 12–17, 1994. It was organized by Stuart Hameroff, an anesthesiologist at the University of Arizona whose work on microtubules had influenced Roger Penrose's view of the role of quantum effects in consciousness. The meeting was thus dominated by speakers from the quantum-consciousness school of neuroscience. They included not only Roger Penrose but also Brian Josephson, a Nobel laureate in physics who has suggested that quantum effects can explain mystical and even psychic phenomena; Andrew Weil, a physician and authority on psychedelia who has asserted that a complete theory of the mind must take into account the ability of South American Indians who have ingested psychotropic drugs to experience shared hallucinations; and Danah Zohar, a New Age author who has declared that human thought stems from "quantum fluctuations of the vacuum energy of the universe," which "is really God." I described this meeting

in "Can Science Explain Consciousness?" *Scientific American*, July 1994, pp. 88–94.

21. David Chalmers set forth his theory of consciousness in *Scientific American*, December 1995, pp. 80–86. In an accompanying article, Francis Crick and Christof Koch offered a rebuttal.

22. I interviewed Minsky at MIT in May 1993. During the interview, Minsky confirmed that in 1966 he had told an undergraduate student, Gerald Sussman, to design, as a summer project, a machine that could recognize objects, or "see." Needless to say, Sussman did not succeed (although he did go on to become a professor at MIT). Artificial vision remains one of the most profoundly difficult problems in artificial intelligence. For a critical look at artificial intelligence, see *AI: The Tumultuous History of the Search for Artificial Intelligence,* by Daniel Crevier, Basic Books, New York, 1993. See also Jeremy Bernstein's respectful profile of Minsky in the *New Yorker*, December 14, 1981, p. 50.

23. *The Society of Mind*, Marvin Minsky, Simon and Schuster, New York, 1985. The book is peppered with remarks that reveal Minsky's ambivalence about the consequences of scientific progress. See, for example, the essay on page 68 titled "Self-Knowledge Is Dangerous," in which Minsky declares: "If we could deliberately seize control of our pleasure systems, we could reproduce the pleasure of success without the need for any actual accomplishment. And that would be the end of everything." Gunther Stent predicted that this type of neural stimulation would be rampant in the new Polynesia.

24. Stent, *Golden Age*, pp. 73–74.

25. Ibid., p. 74.

26. Gilbert Ryle coined the phrase "ghost in the machine" in his classic attack on dualism, *The Concept of Mind*, Hutchinson, London, 1949.

27. Henry Adams made this reference to Francis Bacon's materialistic outlook in *The Education of Henry Adams*, p. 484 (see note 3 to Chapter 1). According to Adams, Bacon "urged society to lay aside the idea of evolving the universe from a thought, and to try evolving thought from the universe."

Chapter Eight: The End of Chaoplexity

1. The University of Illinois issued this press release on Mayer-Kress in November 1993. My story on Mayer-Kress's simulation of Star Wars,

titled "Nonlinear Thinking," ran in *Scientific American*, June 1989, pp. 26–28. See also "Chaos in the International Arms Race," by Mayer-Kress and Siegfried Grossman, *Nature*, February 23, 1989, pp. 701–704.

2. Books that have followed in the wake of James Gleick's *Chaos: Making a New Science*, Penguin Books, New York, 1987, and show signs of its influence include *Complexity: The Emerging Science at the Edge of Order and Chaos*, M. Mitchell Waldrop, Simon and Schuster, New York, 1992; *Complexity: Life at the Edge of Chaos*, Roger Lewin, Macmillan, New York, 1992; *Artificial Life: A Report from the Frontier Where Computers Meet Biology*, Steven Levy, Vintage, New York, 1992; *Complexification: Explaining a Paradoxical World through the Science of Surprise*, John Casti, HarperCollins, New York, 1994; *The Collapse of Chaos: Discovering Simplicity in a Complex World*, Jack Cohen and Ian Stewart, Viking, New York, 1994; and *Frontiers of Complexity: The Search for Order in a Chaotic World*, Peter Coveny and Roger Highfield, Fawcett Columbine, New York, 1995. This final book covers much of the material covered by Gleick in *Chaos*, corroborating my point that popular treatments of chaos and complexity have virtually erased the distinction between them.

3. Gleick printed this quote from Poincaré in *Chaos*, p. 321.

4. *The Dreams of Reason*, Heinz Pagels, Simon and Schuster, New York, 1988. I quoted the blurb on the July 1989 paperback edition by Bantam.

5. *The Fractal Geometry of Nature*, Benoit Mandelbrot, W. H. Freeman, San Francisco, 1977, p. 423. The remark earlier in this paragraph that the Mandelbrot set is "the most complex object in mathematics" was made by the computer scientist A. K. Dewdney in *Scientific American*, August 1985, p. 16.

6. I watched Epstein demonstrate his artificial-society program during a workshop held at the Santa Fe Institute in May 1994 (which I describe in detail in the next chapter). I heard Epstein claim that computer models such as his would revolutionize social science during a one-day symposium at the Santa Fe Institute on March 11, 1995.

7. Holland made this claim in an unpublished paper that he sent me, titled "Objectives, Rough Definitions, and Speculations for Echo-Class Models." (The term *Echo* refers to Holland's major class of genetic algorithms.) He reiterated the claim on page 4 of his book, *Hidden Order: How Adaptation Builds Complexity*, Addison-Wesley, Reading, Mass., 1995. Holland presented a succinct description of genetic algorithms in *Scientific American*, July 1992, pp. 66–72.

8. Yorke made this remark during a telephone interview in March 1995. Gleick's *Chaos* credited Yorke with having coined the term *chaos* in 1975.

9. See note 2.

10. See "Revisiting the Edge of Chaos," by Melanie Mitchell, James Crutchfield, and Peter Hraber, Santa Fe working paper 93-03-014. Coveny and Highfield's *Frontiers of Complexity*, cited in note 2, also mentioned the criticism of the edge-of-chaos concept.

11. At this writing, Seth Lloyd still had not published all his definitions of complexity. After I called him to ask about the definitions, he E-mailed the following list, which by my count includes not 31 definitions but 45. The names that are used as modifiers or in parentheses refer to the main originators of the definition. For what it's worth, here is Lloyd's list, only slightly edited: information (Shannon); entropy (Gibbs, Boltzman); algorithmic complexity; algorithmic information content (Chaitin, Solomonoff, Kolmogorov); Fisher information; Renyi entropy; Self-delimiting code length (Huffman, Shannon-Fano); error-correcting code length (Hamming); Chernoff information; minimum description length (Rissanen); number of parameters, or degrees of freedom, or dimensions; Lempel–Ziv complexity; mutual information, or channel capacity; algorithmic mutual information; correlation; stored information (Shaw); conditional information; conditional algorithmic information content; metric entropy; fractal dimension; self-similarity; stochastic complexity (Rissanen); sophistication (Koppel, Atlan); topological machine size (Crutchfield); effective or ideal complexity (Gell-Mann); hierarchical complexity (Simon); tree subgraph diversity (Huberman, Hogg); homogeneous complexity (Teich, Mahler); time computational complexity; space computational complexity; information-based complexity (Traub); logical depth (Bennett); thermodynamic depth (Lloyd, Pagels); grammatical complexity (position in Chomsky hierarchy); Kullbach–Liebler information; distinguishability (Wooters, Caves, Fisher); Fisher distance; discriminability (Zee); information distance (Shannon); algorithmic information distance (Zurek); Hamming distance; long-range order; self-organization; complex adaptive systems; edge of chaos.

12. See chapter 3 of *The Quark and the Jaguar*, W. H. Freeman, New York, 1994, in which Murray Gell-Mann, a Nobel laureate in physics and one of the founders of the Santa Fe Institute, described Gregory Chaitin's

algorithmic information theory and other approaches to complexity. Gell-Mann acknowledged on page 33 that "any definition of complexity is necessarily context-dependent, even subjective."

13. The conference on artificial life held at Los Alamos in 1987 was vividly described in Steven Levy's *Artificial Life*, cited in note 2.

14. Editor's introduction, Christopher Langton, *Artificial Life*, vol. 1, no. 1, 1994, p. vii.

15. See note 2 for full citations.

16. "Verification, Validation, and Confirmation of Numerical Models in the Earth Sciences," by Naomi Oreskes, Kenneth Belitz, and Kristin Shrader Frechette, was published in *Science*, February 4, 1994, pp. 641–646. See also the letters reacting to the article, which were published on April 15, 1994.

17. Ernst Mayr discussed the inevitable imprecision of biology in *Toward a New Philosophy of Biology*, Harvard University Press, Cambridge, 1988. See in particular the chapter entitled "Cause and Effect in Biology."

18. I interviewed Bak in New York City in August 1994. For an introduction to Bak's work, see "Self-Organized Criticality," *Scientific American*, by Bak and Kan Chen, January 1991, pp. 46–53.

19. *Earth in the Balance*, Al Gore, Houghton Mifflin, New York, 1992, p. 363.

20. See "Instabilities in a Sandpile," by Sidney R. Nagel, *Reviews of Modern Physics*, vol. 84, no. 1, January 1992, pp. 321–325.

21. A discussion of Leibniz's belief in an "irrefutable calculus" that could solve all problems, even theological ones, can be found in the excellent book *Pi in the Sky*, by John Barrow, Oxford University Press, New York, 1992, pp. 127–129.

22. *Cybernetics*, by Norbert Wiener, was published in 1948 by John Wiley and Sons, New York.

23. John R. Pierce made this comment about cybernetics on page 210 of his book *An Introduction to Information Theory*, Dover, New York, 1980 (originally published in 1961).

24. Claude Shannon's paper, "A Mathematical Theory of Communications," was published in the *Bell System Technical Journal*, July and October, 1948.

25. I interviewed Shannon at his home in Winchester, Mass., in November 1989. I also wrote a profile of him for *Scientific American*, January 1990, pp. 22–22b.

26. This glowing review of Thom's book appeared in the *London Times Higher Education Supplement*, November 30, 1973. I found the reference in *Searching for Certainty*, by John Casti, William Morrow, New York, 1990, pp. 63–64. Casti, who has written a number of excellent books on mathematics-related topics, is associated with the Santa Fe Institute. The English translation of Thom's book *Structural Stability and Morphogenesis*, originally issued in French in 1972, was published in 1975 by Addison-Wesley, Reading, Mass.

27. These negative comments about catastrophe theory were reprinted in Casti, *Searching for Certainty*, p. 417.

28. *Chance and Chaos*, David Ruelle, Princeton University Press, Princeton, N.J., 1991, p. 72. This book is a quiet but profound meditation on the meaning of chaos by one of its pioneers.

29. "More Is Different," Philip Anderson, *Science*, August 4, 1972, p. 393. This essay is reprinted in a collection of papers by Anderson, *A Career in Theoretical Physics*, World Scientific, River Edge, N.J., 1994. I interviewed Anderson at Princeton in August 1994.

30. As quoted in "The Man Who Knows Everything," by David Berreby, *New York Times Magazine*, May 8, 1994, p. 26.

31. I first described my encounter with Gell-Mann in New York City, which took place in November 1991, in *Scientific American*, March 1992, pp. 30–32. I interviewed Gell-Mann at the Santa Fe Institute in March 1995.

32. See note 12.

33. "Welcome to Cyberia: Notes on the Anthropology of Cyberculture," Arturo Escobar, *Current Anthropology*, vol. 35, no. 3, June 1994, p. 222.

34. *Order out of Chaos*, Ilya Prigogine and Isabelle Stengers, Bantam, New York, 1984 (originally published in French in 1979).

35. Ibid., 299–300.

36. Ruelle, *Chance and Chaos*, p. 67.

37. Feigenbaum's two papers were "Presentation Functions, Fixed Points, and a Theory of Scaling Function Dynamics," *Journal of Statistical Physics*, vol. 52, nos. 3/4, August 1988, pp. 527–569; and "Presentation Functions and Scaling Function Theory for Circle Maps," *Nonlinearity*, vol. 1, 1988, pp. 577–602.

Chapter Nine: The End of Limitology

1. The meeting titled "The Limits to Scientific Knowledge" was held May 24–26, 1994, at the Santa Fe Institute.

2. See Chaitin's articles "Randomness in Arithmetic," *Scientific American*, July 1988, pp. 80–85; and "Randomness and Complexity in Pure Mathematics," *International Journal of Bifurcation and Chaos*, vol. 4, no. 1, 1994, pp. 3–15. Chaitin has also distributed a book called *The Limits of Mathematics* on the Internet. Other relevant publications by participants at this meeting (listed alphabetically by author, and not including publications already cited) are W. Brian Arthur, "Positive Feedbacks in the Economy," *Scientific American*, February 1990, pp. 92–99; John Casti, *Complexification*, HarperCollins, New York, 1994; Ralph Gomory, "The Known, the Unknown, and the Unknowable," *Scientific American*, June 1995, p. 120; Rolf Landauer, "Computation: A Fundamental Physical View," *Physica Scripta*, vol. 35, pp. 88–95, and "Information Is Physical," *Physics Today*, May 1991, pp. 23–29; Otto Rossler, "Endophysics," *Real Brains, Artificial Minds*, edited by John Casti and A. Karlqvist, North Holland, New York, 1987, pp. 25–46; Roger Shepard, "Perceptual-Cognitive Universals as Reflections of the World," *Psychonomic Bulletin and Review*, vol. 1, no. 1, 1994, pp. 2–28; Patrick Suppes, "Explaining the Unpredictable," *Erkenntis*, vol. 22, 1985, pp. 187–195; Joseph Traub, "Breaking Intractability" (cowritten with Henry Wozniakowski), *Scientific American*, January 1994, pp. 102–107. I discussed some of the mathematics-related topics that arose at the Santa Fe meeting in "The Death of Proof," *Scientific American*, October 1993, pp. 92–103. One of the best books I have read recently on the limits of knowledge is *Fire in the Mind*, by George Johnson, Alfred A. Knopf, New York, 1995.

3. I interviewed Chaitin on the Hudson River in September 1994.

4. *The End of History and The Last Man*, Francis Fukuyama, The Free Press, 1992.

5. I wrote a profile of Brian Josephson for *Scientific American*, "Josephson's Inner Junction," May 1995, pp. 40–41.

Chapter Ten: Scientific Theology, or The End of Machine Science

1. *The World, the Flesh and the Devil*, J. D. Bernal, Indiana University Press, Bloomington, 1929, p. 47. I am indebted to Robert Jastrow of Dartmouth College for sending me a copy of Bernal's essay.

2. *Mind Children*, Hans Moravec, Harvard University Press, Cambridge, 1988. My interview with Moravec took place in December 1993.

3. *Infinite in All Directions*, Freeman Dyson, Harper and Row, New York, 1988, p. 298.

4. Dyson's romantic view of science brings him to the verge of radical relativism. See his essay "The Scientist as Rebel," *New York Review of Books*, May 25, 1995, p. 31.

5. "Time without End: Physics and Biology in an Open Universe," Freeman Dyson, *Reviews of Modern Physics*, vol. 51, 1979, pp. 447–460.

6. Dyson, *Infinite in All Directions*, p. 196.

7. Ibid., p. 115.

8. Ibid., pp. 118–119.

9. The fascinating career of Edward Fredkin (as well as those of Edward Wilson and the late economist Kenneth Boulding) is described in *Three Scientists and Their Gods*, by Robert Wright, Pantheon, New York, 1988. I interviewed Fredkin by telephone in May 1993.

10. *The Physics of Immortality*, Frank Tipler, Doubleday, New York, 1994.

11. *The Anthropic Cosmological Principle*, Frank Tipler and John Barrow, Oxford University Press, New York, 1986.

12. Teilhard de Chardin discussed the question of how extraterrestrials might be saved in the chapter titled "A Sequel to the Problem of Human Origins: The Plurality of Inhabited Worlds," in *Christianity and Evolution*, Harcourt Brace Jovanovich, New York, 1969.

13. *The Limits of Science*, Peter Medawar, Oxford University Press, New York, 1984, p. 90.

Epilogue: The Terror of God

1. See "The Mystery at the End of the Universe," chapter 9 of *The Mind of God*, by Paul C. Davies, Simon and Schuster, New York, 1992.

2. See "Borges and I," *A Personal Anthology*, by Jorge Luis Borges, Grove Press, New York, 1967, pp. 200–201.

3. See note 9 of the chapter titled "Mysticism" in *The Varieties of Religious Experience*, by William James, Macmillan, New York, 1961 (James's book was originally published in 1902).

4. See *The Philosophy of Charles Hartshorne*, edited by Lewis Edwin Hahn, Library of Living Philosophers, La Salle, Ill., 1991. I spoke to Hartshorne in May 1993.

5. *Tractatus Logico-Philosophicus*, Ludwig Wittgenstein, Routledge, New York, 1990 edition, p. 187. Wittgenstein's cryptic book was originally published in 1922.

Selected Bibliography

Barrow, John, *Theories of Everything*, Clarendon Press, Oxford, U.K., 1991.

Barrow, John, *Pi in the Sky*, Oxford University Press, New York, 1992.

Bloom, Harold, *The Anxiety of Influence*, Oxford University Press, New York, 1973.

Bohm, David, *Wholeness and the Implicate Order*, Routledge, New York, 1980.

Bohm, David, and F. David Peat, *Science, Order and Creativity*, Bantam Books, New York, 1987.

Borges, Jorge Luis, *A Personal Anthology*, Grove Press, New York, 1967.

Casti, John, *Searching for Certainty*, William Morrow, New York, 1990.

Casti, John, *Complexification*, HarperCollins, New York, 1994.

Chomsky, Noam, *Language and the Problems of Knowledge*, MIT Press, Cambridge, Mass., 1988.

Coveny, Peter, and Roger Highfield, *Frontiers of Complexity*, Fawcett Columbine, New York, 1995.

Crick, Francis, *Life Itself*, Simon and Schuster, New York, 1981.

Crick, Francis, *What Mad Pursuit*, Basic Books, New York, 1988.

Crick, Francis, *The Astonishing Hypothesis*, Charles Scribner's Sons, New York, 1994.

Davies, Paul C., *The Mind of God*, Simon and Schuster, New York, 1992.

Dawkins, Richard, *The Blind Watchmaker*, W. W. Norton, New York, 1986.

Dennett, Daniel, *Consciousness Explained*, Little, Brown, Boston, 1991.

Dyson, Freeman, *Infinite in All Directions*, Harper and Row, New York, 1988.

Edelman, Gerald, *Neural Darwinism*, Basic Books, New York, 1987.

Edelman, Gerald, *Topobiology*, Basic Books, New York, 1988.

Edelman, Gerald, *The Remembered Present*, Basic Books, New York, 1989.

Edelman, Gerald, *Bright Air, Brilliant Fire*, Basic Books, New York, 1992.

Feyerabend, Paul, *Against Method*, Verso, London, 1975.

Feyerabend, Paul, *Farewell to Reason*, Verso, London, 1987.

Feyerabend, Paul, *Killing Time*, University of Chicago Press, Chicago, 1995.

Fukuyama, Francis, *The End of History and The Last Man*, The Free Press, New York, 1992.

Geertz, Clifford, *The Interpretation of Cultures*, Basic Books, New York, 1973.

Geertz, Clifford, *Works and Lives*, Stanford University Press, Stanford, Calif., 1988.

Geertz, Clifford, *After the Fact*, Harvard University Press, Cambridge, Mass., 1995.

Gell-Mann, Murray, *The Quark and the Jaguar*, W. H. Freeman, New York, 1994.

Gleick, James, *Chaos: Making a New Science*, Penguin Books, New York, 1987.

Gould, Stephen Jay, *Wonderful Life*, W. W. Norton, New York, 1989.

Harwit, Martin, *Cosmic Discovery*, MIT Press, Cambridge, Mass. 1981.

Hawking, Stephen, *A Brief History of Time*, Bantam Books, New York, 1988.

Holton, Gerald, *Science and Anti-Science*, Harvard University Press, Cambridge, Mass., 1993.

Hoyle, Fred, *Home Is Where the Wind Blows*, University Science Books, Mill Valley, Calif., 1994.

Hoyle, Fred, and Chandra Wickramasinghe, *Our Place in the Cosmos*, J. M. Dent, London, 1993.

Johnson, George, *Fire in the Mind*, Knopf, New York, 1995.

Kauffman, Stuart, *The Origins of Order*, Oxford University Press, New York, 1993.

Kauffman, Stuart, *At Home in the Universe*, Oxford University Press, New York, 1995.

Kuhn, Thomas, *The Structure of Scientific Revolutions*, University of Chicago Press, Chicago, 1962.

Levy, Steven, *Artificial Life*, Vintage, New York, 1992.

Lewin, Roger, *Complexity*, Macmillan, New York, 1992.

Lindley, David, *The End of Physics*, Basic Books, New York, 1993.

Mandelbrot, Benoit, *The Fractal Geometry of Nature*, W. H. Freeman, San Francisco, 1977.

Margulis, Lynn, *Symbiosis in Cell Evolution*, W. H. Freeman, New York, 1981.

Margulis, Lynn, and Dorion Sagan, *What Is Life?*, Peter Nevraumont, Inc., New York, 1995.

Mayr, Ernst, *Toward a New Philosophy of Biology*, Harvard University Press, Cambridge, Mass., 1988.

Mayr, Ernst, *One Long Argument*, Harvard University Press, Cambridge, Mass., 1991.

McGinn, Colin, *The Problem of Consciousness*, Blackwell, Cambridge, Mass., 1991.

McGinn, Colin, *Problems in Philosophy*, Blackwell, Cambridge, Mass., 1993.

Minsky, Marvin, *The Society of Mind*, Simon and Schuster, New York, 1985.

Moravec, Hans, *Mind Children*, Harvard University Press, Cambridge, Mass., 1988.

Overbye, Dennis, *Lonely Hearts of the Cosmos*, HarperCollins, New York, 1992.

Pagels, Heinz, *The Dreams of Reason*, Simon and Schuster, New York, 1988.

Penrose, Roger, *The Emperor's New Mind*, Oxford University Press, New York, 1989.

Penrose, Roger, *Shadows of the Mind*, Oxford University Press, New York, 1994.

Popper, Karl, and John C. Eccles, *The Self and its Brain*, Springer-Verlag, Berlin, 1977.

Popper, Karl, *Unended Quest*, Open Court, La Salle, Ill., 1985.

Popper, Karl, *Popper Selections*, edited by David Miller, Princeton University Press, Princeton, N.J., 1985.

Prigogine, Ilya, *From Being to Becoming*, W. H. Freeman, New York, 1980.

Prigogine, Ilya, and Isabelle Stengers, *Order out of Chaos*, Bantam, New York, 1984 (originally published in French in 1979).

Rescher, Nicholas, *Scientific Progress*, Basil Blackwell, Oxford, U.K., 1978.

Rescher, Nicholas, *The Limits of Science*, University of California Press, Berkeley, 1984.

Ruelle, David, *Chance and Chaos*, Princeton University Press, Princeton, N.J., 1991.

Selve, Richard Q., editor, *The End of Science? Attack and Defense*, University Press of America, Lanham, Md., 1992.

Stent, Gunther, *The Coming of the Golden Age*, Natural History Press, Garden City, N.Y., 1969.

Stent, Gunther, *The Paradoxes of Progress*, W. H. Freeman, San Francisco, 1978.

Tipler, Frank, *The Physics of Immortality*, Doubleday, New York, 1994.

Tipler, Frank, and John Barrow, *The Anthropic Cosmological Principle*, Oxford University Press, New York, 1986.

Waldrop, Mitchell, *Complexity*, Simon and Schuster, New York, 1992.

Weinberg, Steven, *Dreams of a Final Theory*, Pantheon, New York, 1992.

Wheeler, John, and Wojciech H. Zurek, editors, *Quantum Theory and Measurement*, Princeton University Press, Princeton, N.J., 1983.

Wheeler, John, *At Home in the Universe*, American Institute of Physics Press, Woodbury, N.Y., 1994.

Wilson, Edward O., *Sociobiology*, Harvard University Press, Cambridge, Mass., 1975.

Wilson, Edward O., *On Human Nature*, Harvard University Press, Cambridge, Mass., 1978.

Wilson, Edward O., and Charles Lumsden, *Genes, Mind and Culture*, Harvard University Press, Cambridge, Mass., 1981.

Wilson, Edward O., and Charles Lumsden, *Promethean Fire*, Harvard University Press, Cambridge, Mass., 1983.

Wilson, Edward O., *Naturalist*, Island Press, Washington, D.C., 1994.

Wright, Robert, *Three Scientists and Their Gods*, Times Books, New York, 1988.

Wright, Robert, *The Moral Animal*, Pantheon, New York, 1994.

Index

Edelman, Gerald, 165–72, 184, 284nn. 7, 8, and 10

Einstein, Albert, 1, 5, 6, 8, 16–19, 60, 65, 68, 70, 86, 107, 111, 172, 195, 218, 219, 221, 231, 236, 238, 272n. 24

Einstein-Podolsky-Rosen paradox, 86–87, 276n. 22

Eisley, Loren, 279n. 19

Eldredge, Niles, 120–21, 124, 280n. 7

Ellsworth, Henry, 20–21

emergence, 125, 192, 214. See also "More Is Different"

Emperor's New Mind, The (Penrose), 1–2, 174–76, 269n. 1, 284–85nn. 14–15

End of Science?, The, (Gustavus Adolphus symposium), 9, 12, 62, 269n. 1, 275n. 2

End of Physics, The (Lindley), 70, 275n. 7, 278n. 6

Engels, Friedrich, 22–23, 271n. 22

engineering. See science, applied

Epstein, Joshua, 195, 287n. 6

Escobar, Arturo, 216, 290n. 33

ET (movie), 131

evil, 77, 259–60. See also suffering, human

evolutionary biology, 4, 6, 14, 104, 112, 114–42
    social science and, 143–49, 151

evolutionary psychology, 146

extraterrestrial life, 17, 101, 109, 117–18, 126, 131, 133, 141–42, 242, 292n. 12

falsification, 34, 38–41, 43, 47, 52

Farewell to Reason (Feyerabend), 49, 274n. 21

fascism, 34, 41, 49, 54

Feigenbaum, Mitchell, 221–25, 290n. 37

Ferris, Timothy, 269n. 1, 270n. 11

Feyerabend, Paul, 32, 33, 48–56, 57, 59, 74, 150, 188, 274nn. 20–28

Feynman, Richard, 90–91, 175, 240, 253, 270n. 14, 277n. 25

Finnegans Wake (Joyce), 71, 197, 211

Fire in the Mind (Johnson), 278n. 6, 291n. 2

Flanagan, Owen, 178, 285n. 16

Fractal Geometry of Nature, The (Mandelbrot), 194, 287n. 5

fractals, 193–94, 224, 226, 236, 238

Fredkin, Edward, 255–56, 292n. 9

free will, 57, 152, 161, 173, 177. See also dualism; determinism

Freud, Sigmund, 34, 185, 189

Frye, Northrop, 269n. 4

Fukuyama, Francis, 243–44, 250, 255, 291n. 4

fundamentalism, religious, 5, 24, 53, 220

fusion reactors, 11, 17

Gaia, 129–31, 137, 148, 281n. 14

Galileo, 46, 48, 62

Gates, Bill, 271n. 17

Geertz, Clifford, 154–58

Gell-Mann, Murray, 65, 211–15, 226, 288nn. 11–12, 290n. 31

genetic algorithms, 195

genetic engineering, 18, 241, 247, 253

genetics, 10, 57, 115, 118, 144, 146, 159, 162. See also DNA; Mendel; molecular biology; sociobiology

Georgi, Howard, 104–5, 279n. 10

Gibbons, Jack, 72

Glashow, Sheldon, 62–65, 95, 104, 275nn. 2–3

Glass, Bentley, 24–26, 28, 29, 272n. 32

Gleick, James, 192, 193, 204, 221, 287nn. 2–3, 288n. 8

God, 6, 55, 60, 71, 76, 77, 94, 109, 117, 149, 237, 254, 257, 259
    terror of, 262–66
    See also Omega Point; scientific theology

Gödel, Kurt, 5–6, 69, 83

Gödel's incompleteness theorem, 5–6, 69, 95, 174, 175, 177, 215, 227, 228, 230–31, 238, 239–40, 242, 250, 254, 261

Goethe, 135

Gold, Thomas, 106

Gomory, Ralph, 229, 234, 238, 291n. 2

Goodwin, Brian, 135, 281n. 18

Gore, Al, 206, 289n. 19

About the Author

John Horgan is a senior writer at *Scientific American*, where he has worked since 1986. He has won the American Association for the Advancement of Science Journalism Award (twice) and the National Association of Science Writers Science-in-Society Award. His articles have appeared in the *New York Times Book Review*, *New Republic*, *Discover*, *New Scientist*, *Science*, and *Omni*. He graduated from Columbia University's School of Journalism in 1983. He lives in upstate New York with his wife, Suzie Gilbert, who is also an author, and their two children.